From Darwin to Eden

From Darwin to Eden

A Tour of Science and Religion
based on the Philosophy of Michael Polanyi
and the Intelligent Design Movement

WILLIAM B. COLLIER

RESOURCE *Publications* · Eugene, Oregon

FROM DARWIN TO EDEN
A Tour of Science and Religion based on the Philosophy of Michael Polanyi and
the Intelligent Design Movement

Resource Publications
An Imprint of Wipf and Stock Publishers
199 W. 8th Ave., Suite 3
Eugene, OR 97401

www.wipfandstock.com

PAPERBACK ISBN: 978-1-5326-9271-0
HARDCOVER ISBN: 978-1-5326-9272-7
EBOOK ISBN: 978-1-5326-9273-4

Manufactured in the U.S.A.   03/04/20

To Susan, Jessica, Justin, Andrew, and Tiffany

Contents

Preface

"BILL, MOST OF THE people I meet with are not concerned much with origin and scientific apologetics, it is just not an issue of importance to them."

That was the response I got from one pastor when informed of the book and its contents. Interestingly, another pastor had me do an eight week Wednesday night series for the church on the book topics. Another pastor teamed up with me and a church elder to present several public talks on the philosophical and scientific case for God. Another pastor and the church elders invited me to give a Sunday morning sermon to the church on these topics that was surprisingly well received. So who is right? I would argue all are. Some people just do not think about these things, have it all figured out, or just do not care. But there are a lot of people, who are very interested, curious, or just puzzled over it. That group is a lot bigger than most people realize, especially among young adults, who are trying to figure out who they are, what they believe, and where they are going in life. To those who are studying for a career in science and other areas; discussion of these issues can be a critical factor in how they spend their lives. I have seen dramatic changes in viewpoints and life direction after exposure to this material. It wasn't me, it was the material.

This book comes out of many years of teaching university science students and half of it co-teaching a course on the interaction of science and faith. My fellow co-teachers in those years were various theology professors. But the primary co-teacher was Dr. Samuel Thorpe, present chair of the undergraduate theology program. Sam was instrumental in helping me develop a philosophical outlook on the problem, and how you teach honor students philosophical material. His imprint is all over this book and without his influence it would have been a different book and much poorer.

I entered graduate school in physical chemistry wondering if my faith would take a serious beating with graduate level instruction. To my surprise, it was not my faith that took a beating, but certain perceptions I had about science and its ability to find absolute truth. Nothing impresses a young mind quite like learning the foundational postulates of a discipline; and that is one thing graduate school teaches with a vengeance. My graduate statistical thermodynamics class used Shannon's information theory to derive the Boltzmann equation. The issue of information in DNA kept haunting me. There was something significant about this. A decade later I found out I was not the only one thinking along those lines.

On a research sabbatical to Budapest Hungary, I stumbled on the works of Michael Polanyi. Polanyi was a famous internationally known Hungarian scientist with profound things to say about how science works. As the wife of Michael Polanyi's son told me, "Everybody reads him!" Because he was a physical chemist like me, I found a closer kinship with his experiences and writing, and his science philosophy closely matched what I have experienced in science. In my opinion, his *Science, Faith and Society* text remains one of the most under-appreciated masterpieces of the entire science and religion genre. It is the best and most accurate description of how science works, that I have ever read. I stumbled upon Michael Behe's *Darwin's Black Box* during this time, and thus was born a merger between the intelligent design movement and the philosophy of Michael Polanyi that continues to this day. The reaction of my students and others has been gratifying. Typically it is one of amazement, that they had not been exposed to this material earlier. I share their sentiments. Whether you agree with it or not, all serious philosophical, theological students, scholars and interested layman need to be aware of this information. To fully grasp the modern intelligent design movement requires reading many quite technical books. To my surprise few have attempted a synopsis that is not trivial, can be covered in one book, at a level that the layman can read, and with which the expert is not bored. I have tried to bind it together with the insightful philosophy of Michael Polanyi. Why? Many claim Polanyi invented intelligent design. Few positivists and others of that thinking, attempt rebuttal of Polanyi's philosophy; they are likely too scared to try, don't understand its ramifications, or are clueless on how to do it. So the most common approach among his detractors is just to ignore him or misrepresent what he really said. Unfortunately the depth of his thought has not made his books accessible to people who would be greatly encouraged by what he had to say. That is why I have used him as my steering philosophy in this book. Reading Polanyi can change your life. I have tried to be broad in this book. There is no promotion of a particular sect, or denominational doctrine. Thus hopefully the book can be used by a variety of readers and teachers, be they secular, theistic, or Christian. I address in particular theism, materialism, and what C. S. Lewis calls "mere Christianity," and how it interacts with modern science.

Several literary devices are used to communicate the ideas as clearly as I can and to make it interesting for the reader. I have technical discussion of the ideas, hopefully at a level a bright high school student, typical college student or interested layman can follow. Over the years I have experienced interesting and sometimes unusual conversations with others on science and religion topics. Some are informative and some quite humorous. These are included to make it interesting and communicate finer points. Chapter five covers the minimal basics of organic chemistry, biochemistry, and molecular biology at a layman's level. If you have only had high school chemistry, and hated it, this chapter is for you. I assume as little as possible and put in interesting conversations to see if I can communicate the smallest amount of needed essentials of three years of chemistry to the non-scientist in a graphical and fun fashion. Time will tell if I succeeded. To the chemical specialist, please ignore the gross simplifications used, read the first part on the cell being self-reproducing automata, and skip the rest of chapter five. The purpose is to help the non-scientist understand the significance of the material presented in chapters six Primordial Soup, and chapter eight Change and Evolution. To the non-scientist, if you get bogged down in chapter five, do not worry about it, skip over it, get as much as you can out of chapters six and eight, and the rest of the book will still be very readable. Everybody loves a romance, so I put one in the book. The conversation between male and female students is entirely made up and has no similarity to people and situations that I know. Any similarity is entirely coincidental. But these conversations also serve to inject a bit of humor into what is traditionally a serious subject. They also serve as a platform for reinforcing new ideas by repeating them in a different manner. They also speak into the human condition. I wondered if it was wise to include these conversations until after a semester class where I was testing the half-finished manuscript. A young co-ed student came to me and asked, "Do Jon and Ellie ever get together?" She was intrigued. In other places where I am describing my experiences, they are real and the conversations are accurate to the extent I remember them, which is only in substance.

Most of the material presented in this book was presented by many others in more technical publications; Stephen Meyers, Hugh Ross, Fez Rana, Michael Behe, Phillip Johnson, Casey Luskin, Douglas Axe, Charles Thaxton, Walter Bradley, Roger Olsen, Alan Chambers, Michael Denton, Fred Heeren, Lori Kanitz, Zeb Long, Douglas Murray, Jay Richards, Guillermo Gonzalez, David Lindberg, Allen Debus, Thomas Kuhn, Alan Chambers, Alister McGrath, Gerald Rau, Janet Soskice, J. William Schopf, Laura Synder, Duane Thurman, John West, Jeffrey Russell, Martin Moleski, and lastly the deceased but not forgotten Michael Polanyi. I strongly encourage all interested readers to read further into these fine scholars. To that cause I have referenced and documented their books and papers extensively in this manuscript. Some of these scholars will agree with some of my conclusions, others will disagree with my conclusions, a few completely. A small number of these scholars I have seen or met at conferences, and a very few I have had the chance to bring to my university to speak. To all I am very grateful.

Acknowledgements

First thanks go to my family, especially my wife Susan. As my now and forever love, she possess a very keen mind that catches my grammar, sloppy logic, and occasional lousy attitude. Without her support, life-long banter, and thought; this book would never have happened. Few women would allow their husband to drag them and their four young kids to the central European metropolis of Budapest, Hungary to live. The first time we lived there, Hungary was just six years out of communism, and scrambling to rebuild their economy. The people of Hungary and the computational chemistry faculty of Eotvos Lorand University, Budapest were our best friends, and destroyed any fears I had of being the ugly American. The genesis of this book had its origin during our second year-long stay there. There I had the time to read and delve into the writings of Michael Polanyi. He was on track for a Nobel Prize in chemistry, but his diversion into personal knowledge probably destroyed his chances. My children have all endured Dad pontificating on the themes in this book, and my oldest daughter Jessica took the class on science and religion from me and my co-teacher Dr. Thorpe. Her analysis and thoughts as a student were very helpful. My youngest daughter Tiffany was responsible for drawing many of the diagrams in this book. If the diagram has a free-hand curve in it, she drew it. I am particularly fond of the reptile to mammal skull transition sequence that she drew.

Several professors, scholars, friends and co-workers have presented guest lectures in our class. Their input was invaluable and weaves it way into this book. In particular, thanks goes to Lori Kanitz, whose science and metaphor lecture merited a chapter in this book with her kind permission; Ken Weed for the cytochrome C diagram and thoughts; Sam Thorpe for the philosophy introduction; Don Vance for lectures on Genesis that were unforgettable; Hal Reed for educating on biological evolution; Andrew Lang on astronomy and relativity, Dominic Halsmer on affordances in optimal design, Gyle Smith on Christian epistemology, Jon Bartlett, Chris Tiews, Nate Maleen, Jeff Barbeau, Steven Herr, Bill Ranahan, and the many other lecturers I have forgotten to mention. To the eighteen years of Honors 102 students that tested and learned this material goes the greatest thanks, and deepest admiration. Our world is in good hands if they are representative of our future leaders. I thank my university for supporting the honors class, its subsequent off-shoots, and the conferences I have attended and sponsored.

Lastly may I thank the excellent staff and editors of Wipf and Stock Publishers, who agreed to take this project on.

1

Introduction:
The World was Flat

"DR. COLLIER? MAY I speak to the class next class period regarding the film we have just seen? I don't believe it is true and I want to explain to the class why it isn't." This was not the sort of request I wanted to hear; particularly coming from a very bright Hungarian Chemistry graduate student; and particularly at the start of my Philosophy and History of Chemistry graduate seminar. My family and I were five months into my year-long sabbatical stay at one of Hungary's finest chemistry programs at their leading university. Through a long series of unusual events I found myself teaching this weekly graduate seminar to a small class of Hungarian graduate students. I was a chemist, not a historian or philosopher, and so were they. Was it the blind leading the blind or did I have anything significant to say to them? I had just shown them a movie called *The Shape of the Earth* made by two History of Science professors back in the States. My student was not happy. "The medieval period did believe the world was flat, and I want to show the class why!" I replied, "This movie was made by professional historians who know what they are talking about, are you sure you want to take them on?" We agreed that he would have his chance next class period and I went home worried about what kind of Pandora's Box I had opened.

At this point your mind may be racing with, "Wait a minute! What do you mean the dark ages knew the world was round and not flat? Didn't Columbus sail the ocean blue to prove it was round in 1492?" If so, welcome to the crowd. The vast majority of modern western civilization, including many modern scientists; and surprisingly, even some historians believe the medieval period was dark, ignorant, and that they believed the world was flat. Daniel Boorstin, a former Librarian of Congress wrote in a popular book called *The Discoverers:*

> A European-wide phenomenon of scholarly amnesia . . . afflicted the continent from AD 300 to at least 1300. During those centuries Christian faith and dogma suppressed the useful image of the world that had been so slowly, so painfully and so scrupulously drawn by ancient geographers.[1]

In a later chapter entitled "A Flat Earth Returns[2]," Boorstin derides the Christian geographers who plunged the world once again into the flat earth belief. There is one small problem with this. Medievals never believed in a flat earth except for one or two ignored eccentrics. Jeffrey Burton Russell, Professor Emeritus of Medieval History at the University of California, Santa Barbara wrote a fascinating book called *Inventing the Flat Earth Columbus and Modern Historians*[3]. His written tweak on modern culture examines in detail how the medieval period believed the world was spherical. The myth of a "Dark Ages" flat earth belief was invented around 1870, established in modern thought through the 1930s, and has persisted in varying degrees until today. It is fairly easy to disprove the flat earth myth. Find a medieval era statue of a king holding the symbol of the divine right of kings in his hand. It will be a "globe" with a cross on top. The globe represents the world as they understood it, and the cross God's divine right of rule given to the King to rule the "world." Next, read the late medieval bestseller book, *Dante's Inferno*. Here as the reader descends through various levels of hell, we encounter Satan at the very last level, in the center of the earth, where he is neither up nor down as there is no such state in the center of the earth. Dante did not even feel compelled to even explain his spherical worldview to his reader, they already knew it.

Then why were some of us taught that the medieval period believed that the world was flat? What else have I not been told about? This is a very good question. Russell's book gives a history of how the flat earth myth happened and some good guesses as to why; chronological snobbery, philosophical worldview agendas, the power of a popular fictional book to establish truth in the mind of the reader, etc. There are lots of possible reasons and they

---

1. Boorstin, *The Discoverers*, 100.

2. Boorstin, *The Discoverers*, 107.

3. Russell, *Inventing the Flat Earth*.

will all have some role to play in our discussion of the philosophy of science and how it affects our societal view on who we are and how we came to be.

Every society, at every age of man, has had to deal with its own very ethnocentric view of itself, and some self-imposed myths regarding how the world works and came to be. They can be found in very surprising places and sometimes be difficult to dislodge. The Christian view of humanity gives a succinct view of why this is so—we are sinners and really hung up on ourselves and our own culture. But how does this play into modern science and philosophy? Good question, let's find out.

Oh, what happened to the Hungarian student? He showed up in class the next week and did not mention a thing. I asked him after class why he didn't bring the topic up again. "I checked it out on the internet and found out you were right," he said. Sigh.

2

Philosophy First

"PLATO!" THE STUDENT EXCLAIMED, "Someone is messing around with my wife! I have clues as to who it may be!"

"Wait" replied Plato, "Examine yourself, are your motives pure in this matter?"

The student replied, "Well, I am pretty mad about this."

"Is your searching out of this problem noble?" the teacher asked.

"But there is nothing noble about messing around with my wife!"

"Will this advance truth in the world?" the great instructor asked.

"I guess not." mumbled the frustrated student.

"Then why are we pursuing this conversation?"

"I see; you are right; sorry that I bothered you about such a trivial matter." The student sadly turned to aimlessly wander off.

"Remember my friend, if a matter is not noble, true, and worthy of consideration it benefits no one." said the great one.

The student turned gazed at Plato fully and said, "Thank you my teacher" and then walked away to ponder greater issues. Plato quietly turned to resume his lecture, relieved that he had not been found out.

While this fictional story is amusing, it does express the suspicions many people have regarding philosophy. But such thoughts need to be counter-balanced with the concept that there is no such thing as a philosophically neutral or totally objective worldview. Like it or not, we perceive this world with a particular viewpoint that is unique to us, our culture, and our times. Am I advocating a relativistic approach to life and philosophy? Absolutely not. But truth is hidden in our worldviews, and we must be

1

careful in sorting through our presuppositions if we are going to find it. A brief look at important philosophical concepts can clear away confusion in controversial issues because it helps us see the lens through which we view the world. It helps us to better analyze the ideas of others.

What is philosophy? The Greek meaning of the word is love of wisdom. A modern definition would say that philosophy is concerned with the ultimate questions and the ultimate answers. Some would argue that true philosophy is the search for truth. You could break the disciplines of philosophy into the following areas. 1—Truth . . . Epistemology or how do we come to know? 2—Reality . . . Metaphysics or what is the nature of the world? 3—Value . . . Axiology or what are values and where did they come from? Before we finish with this book we will find ourselves dipping into many of these areas. Let us look briefly at the first area where it can help us understand the origins issue better.

Epistemology, how do we know anything? What are some of the ways we come to know? Some philosophers have compiled a list like this.

1—empiricism	You come to know through your senses
2—rationalism	Reason and rationality is the way; your mind is better than your senses
3—naturalism	The material world is all there is, so knowledge is found through its investigation
4—logic	You use logic to find truth from your initial presumptions
5—existentialism	You choose your values and way of finding your true knowledge.
6—revelation	Knowledge is revealed through a divine or unknown source

Looking at a list like this for the first time can be very instructive. If you have grown up in a modern western culture it is easy to spot which ways of knowing are viewed favorably and which are not. In popular western culture all of these are used and valued to some extent. In academic circles, revelation knowledge is not regarded as trustworthy. Academia started to rely heavily on empiricism with the Renaissance and Middle Ages.

In spite of popular conception, the rise of empiricism, (knowing and understanding by your senses alone) in modern culture can be traced back to the "so-called" dark ages; or middles ages as modern historians prefer to call them. Allen Bebus[1] in *Man and Nature in the Renaissance* argues that "natural magicians" or investigators of the natural world of that day were slandering the deductive methods of Aristotle and the Peripatic philosophers (those

---

1. Debus, Man and Nature, 11–13.

who followed the philosophy of Socrates, Plato and their students), and advocating a "see and investigate it for yourself" approach to understanding the natural world. With the Scriptures in one hand and your laboratory notebook in the other, one was to pursue natural knowledge by experiments designed and conducted by you. Not by deducing it from the essential presuppositions given by Aristotle and the ancients. The nineteenth century French physicist and historian of science Pierre Duhem argued that modern physics with notions of velocity, acceleration, and graphs were invented a hundred and more years before the Renaissance[2]. Thus many historians will argue that without the initial revelatory knowledge that God created the world, and created it as a separate and distinct non-divine entity from himself, the modern development of the empirical method would not have become what we now call modern science. This freedom to monkey with the physical world around us through empirical methods without offending the gods/God, has its roots in a revealed knowledge. So in a sense modern empiricism was preceded and created by some initial revelatory knowledge. In the case of medieval Europe, the source of revelation knowledge regarding creation was obviously the Bible and Christian theology.

Our culture has tried very hard, but found it very difficult to find true knowledge via empiricism, rationalism, and logic alone. Pearcey and Thaxton[3] have given an interesting glimpse of this journey in their book, *The Soul of Science*. In the 1930s mathematics showed that it was impossible to build a firm provable foundation even for itself. Since mathematics was highly esteemed by the empiricists, the empirical dream for a complete self-sufficient worldview based on knowledge and reason alone, appeared to be futile. First academics, then our popular culture following their lead, turned to existentialism and moved from a "modernist," empiricist rationalist viewpoint to a "post-modern" or more existentialistic viewpoint. The scientific and technical areas of the university system have resisted this change, and have remained more modernist, i.e. empiricist and rationalistic. Their influence on the popular culture and the origins debate is profound. It is strange and peculiar that our western society has embraced existentialism in matters of the heart and values; but become more doggedly empirical in matters of science, origins, and technology. This will be more apparent as we look into the origins issues more closely.

One of the problems of a society moving to a post-modern cultural philosophy is that it becomes very difficult to find a common basis upon which to discuss and solve disagreements. If we cannot agree on what is

2. Lindberg, The Beginnings of Western Science, 295, 357.
3. Pearcey and Thaxton, Soul of Science.

truth)or even on the proper "way" to arrive at truth, the only recourse left to solve major disagreements is raw power. Might makes right. The one with the most power gets to determine what the true and correct way is. It is not a given that our western society is going to be dominated by post-modern power struggles. But if it is, we had better take notice; anarchy has never been fun in any age of history. Any discussion of scientific, technological, and origins issue must deal with the politics and the philosophical background of the culture the debate is taking place in if we are going to understand it. Philosophy gives us tools to help us decide whether we like the direction the debate is going and what we can do to change it, if we do not.

"So Mary what did you bring for lunch?" Jon asked nonchalantly as he eyed her raspberry pie.

"The usual; pita bread, lettuce, tomatoes. Why do you ask?" Mary was sensing that something was afoot, as Jon was not one to idly make conversation about lunch.

"Well do you remember Dr. Swartzmann's lecture on the economics of Soviet cooperation during World War II?"

"Yes, I do," replied Mary suspiciously.

"Remember how under the Soviet system, everyone was encouraged, yes even forced, to collectively share their excess for the common good of the whole community?" Jon intoned.

"Sounds rather brutal to me" murmured Mary.

"But they managed to defeat a vastly superior Nazi army" stated Jon, sensing he was on a roll, "Without a realization that greed can only be suppressed and the common good of the community advanced with an equal sharing of personal resources, they never would have beat the Nazis." Pressing on, Jon said, "Think, Mary what could be accomplished in this country, our very own school, and yes, our classrooms if everyone shared equally of their talents and resources for the common good."

"But what happens if everybody shares and only a few suckers actually do the work necessary to get those resources" asked Mary.

"Come on Mary, the common community will ensure that work is distributed fairly." replied Jon.

"But who in the common community gets to make that decision?" stated Mary.

"Well, all of us of course."

"Oh," replied Mary.

"Mary, I forgot to bring my lunch, how about sharing some of your raspberry pie for the common good?" asked Jon casually. Five minutes later

Jon was in the bathroom cleaning raspberry pie off his face. "I guess she wasn't too interested in the common good," he mumbled.

Let's go back to the list of six ways we come to know and examine each. This will set the stage for a discussion on the philosophy of science, how it has changed over time, and the rather startling place it has come to today. And yes, in case you are wondering, it bears heavily on the science and religion arena.

Empiricism states that we come to know through our senses only, and that this is the most reliable way to know anything. Do I see it, hear it, touch it, feel it, and smell it? It doesn't stop with our simple senses but allows the extension of our senses with instruments that allow us to see and observe where we have never seen before. A microscope shows us the inside of a bacterium. A telescope lets us see other galaxies. The Voyager space probe took cameras and other instruments allowing us to "see" into the rings of Saturn, to visualize the magnetic field surrounding Jupiter. False color photographs show x-ray emission patterns in distance nebula. All of these are but instrumental extensions of our senses, a refined form of empiricism. But is this the only way we come to know things? And how reliable are our senses anyway? How reliable are our instruments? For example, an obvious thought; what is love? We see its effects and actions but it is an abstract idea and concept, and experience. The committed empiricist is tempted to classify love as the specific firing of neurons and synapses in our brain that brings about the experience of love. Try that one on your girlfriend/boyfriend or Mother and see how your loved one responses! What about the experience of values, seeing beauty, despising ugly hateful things; is it just our senses responding? What about the experience of consciousness? I experience you and you experience me in a conversation and are able to uniquely identify each other as separate conscious entities. How is that confined to just empiricism? I think therefore I am. What sense brought that about? It is pretty obvious that empiricism is just part of the picture. A philosophical view of science and knowledge that is heavily based on empiricism, rationalism, and materialism is called positivism. The flaws in this approach will pile up to embarrassing proportions, but amazingly it is still very prevalent in our western culture.

Perhaps we need to be totally logical in our thinking. A logical approach to life will bring us truth. Like a robot who thinks perfectly logically, or the Star Trek Science Officer Spock (when he is not falling in love with a Vulcan woman); if we just applied logical processes to all of our thoughts we can be guaranteed to arrive at failsafe conclusions. The trouble with logic is that it always follows from an initial set of premises. Sometimes we know those premises. Often we don't even realize we are making them as we make our first logical statement. Socrates had a famous conversation with a student where he demonstrated "logically" that the student was a dog. Wrapped up in the initial premises of our logical arguments are the flaws, exceptions, neglected auxiliary hypotheses (will discuss this later) that eventually will expose our wonderful logical arguments as the rubbish they sometimes are. Logic does not produce new knowledge, but rather reveals the knowledge that was hidden in our initial premises. It unravels that piece of latent knowledge that was there all along, but we just could not see it in the confusion of our facts and data. But from where come our initial premises, facts, and data? Empiricism? Common sense observation? Life experiences? Spiritual experiences? Logic is a very useful tool; but to treat it as an all-encompassing way of life is to be very naive about how we really find knowledge. How do you logically show that logic will give you true statements if you start from the right premises? There is an indefinable value judgment wrapped up even in the choice to use logic.

But what about rationalism? Can we solely rely on it as a reliable source of knowledge? Here we bring our minds and thinking into play. Some aspects of knowledge can be thought out. Great theories have come about solely by the process of reasoning and rational thought. It is true that empirical data is used to build a base upon which we think and reason. We must reason and think on this data to come to proper and true conclusions. But there are some catches here. Why would we want to reason to true and correct conclusions? Why should I not rather prefer to reason to a falsehood that gets me and my neighbor killed in a most grievous manner? Maybe pain makes me happy (a common thought of many teachers composing final exams for their students). What is it about rationalism, good, and truthfulness that attract us in such a fashion that we all prefer that route? Values, you say. Try to rationally justify values. It is not easy. Many prefer to just take them as givens; things we just accept as the way life is and move on. Values are necessary preconditions we must accept if we are going to be able to reason and communicate with others. Even the conception of rationality as a truthful and worthy way of finding knowledge is based on a value judgment we cannot rationalize.

Naturalism we earlier defined as, "The material world is all there is, so knowledge is found through its investigation." Another name for naturalism as defined here is philosophical materialism, or just materialism. In this sense we are not talking about the lust for material goods, which can exist in any worldview, but rather a philosophical viewpoint about the nature of the world. Obviously empiricism, rationalism, and logic are tools we use to investigate the material world. But the naturalist or materialist perspective argues that, "that is all" we have available to help us find truth. If it is not materially or rationally conceived, then it just does not exist. There are many flaws to this perspective, but a critical one was clearly stated by the famous Oxford lecturer C. S. Lewis. *Can we trust that our minds can reason correctly?*

> Perhaps this may be even more simply put in another way. Every particular thought (whether is a judgment of thought or a judgment of value) is always and by all men discounted the moment they believed it can be explained, without remainder, as the result of irrational causes. Whenever you know that what the other man is saying is wholly due to his complexes or a bit of bone pressing on his brain you cease to attach importance to it. But if naturalism is true then all thoughts whatever would be wholly due to irrational causes. Therefore all thoughts would be equally worthless. Therefore naturalism is worthless. If it is true, then we can know no truths. It cuts its own throat.[4]

This stabs at the heart of naturalism. If material causes are all that exist, then our own thoughts arise from some process of neurons firing in the brain to cause our thoughts. But there is nothing rational about light photons hitting eyes, causing electrical nerve impulses; that cause brain neurons to fire to cause our thoughts. If there is something above all of this that coordinates these material activities, then it cannot be material by definition.

The Berkeley law professor Phillip Johnson has summarized the naturalist worldview as "the forces and the particles did it all." Here we are using the word naturalist in the philosophical sense of a natural means only or a type of materialism, not as one who admires or learns from nature. Lewis and many others have pointed out that if our minds are simply the result of a long drawn out process of materialistic evolution that needed not a drop of divine intervention or input, then our thoughts themselves are nothing but material processes going on in our brains. We are but machines going about our machine-like business. Thus why should anyone pay attention to the thoughts of one machine over another? Why should we pay attention to any of them? Well, is naturalism completely worthless? By itself as a philosophical worldview it possesses enormous problems. But naturalism as one possible means of investigation (not as an encompassing philosophy), in conjunction with other modes of knowing, can become a very useful tool. We often use natural means to investigate the world and explain many of the phenomena we see there. The investigation of disease and its elimination has been greatly helped by the concept that disease has its origins often in very natural phenomena, rather than in offending the wrong deities and spirits.

4. Lewis, "Religion without Dogma", 144.

These problems with the empirical, logical, or rational worldview have led many in academia to an existential view. If there is no way to find truth as many existentialists argue, then perhaps it doesn't really matter, and we are free to choose truth as it pleases us and makes life good. We choose our values and make them true for our community. However rape and pillage of other men's wives tends to bring wrath upon proponents of such values. Thus those who imbibe of this viewpoint usually temper this with a general respect for others' rights and a need to get along. Self-realization and the maximal development of one's personal self-identity are always hemmed in by its effect on others. Your rights end where my nose begins. But how do you arbitrate such a hemming in of personal rights? Where do you draw the lines of, "Do not cross lest you offend your brother's rights"? Why should I believe you? How are we going to agree on these lines, and who is going to enforce them? Answering these questions will always bring us back to initial presuppositions and our values about what is good and worthwhile in life. You cannot get away from the issue of values and their source.

If all of these modes of knowing have flaws, rationalism, naturalism, logic, existentialism, then what about getting the true knowledge from outside our world or from some inexplicable voice from within us? Let's get true knowledge from revelation—knowledge that is revealed through divine or unknown sources. But how do I know that source is really telling me the truth? How do I know that it even really exists? If we go to supernatural sources, i.e. God, Allah, Krishna, the One, or whatever, then which religion is revealing the truth? It is apparent that more than revelation knowledge will play a role in determining which revelation knowledge source, if any, we will believe; be it a supernatural source outside our universe or a voice within ourselves or the universe itself. If you have several sources of revelation knowledge that give contradictory facts about our origins and existence then somebody is wrong, lying, or something else. So even revelation knowledge will be realized and analyzed within the context of the other forms of knowing. Therefore, we use all of these forms of knowing as we plow through our existence.

The materialist believes the forces and particles in the universe can ultimately explain everything. There is no need for God or anything supernatural to explain our world. They reject revelation knowledge completely or else claim that it can be explained ultimately in terms of the other forms of knowledge. The theist believes that this is not so. There is something, someone, outside the universe who put it all into place who could be that source of revelation knowledge that we often yearn for. Theists come in all different shades, varieties, and religions. I, the writer of this book, am a Christian who, in the words of C. S. Lewis, is a blatant "super-naturalist." I use all of these forms of knowing to explain why I am convinced that Christianity is the truth. This explanation by itself would constitute another book, but the material in this book makes up part of that explanation. So be forewarned. I am not a totally objective writer. I am not sure that anyone can be. That assumes that true objectivity is possible in this world. My Christian perspective argues against that. We do have a sin nature that blinds us to our shortcomings; a nature that put those faults there in the first place. But the Christian perspective also argues that true objectivity can be approached and that the effort to do so is worthwhile and good. Why? Because God is the source of ultimate truth and He can be approached and partially understood by humans.

"Ellie," Charles whispered to her as he lay with his head on her lap looking up at the stars; "Do you think God speaks to people personally?"

"Of course," said Ellie, "Why do you ask?"

"Do you think he could speak to me?" Charles inquired.

"Well, if He speaks to me and other people then he obviously could speak to you," she replied.

"Well, I think he has spoken to me," said Charles.

"Did he speak to you through the Bible or to you personally some way?" she asked.

Charles replied, "Both, aren't they the same? God's word is God's word isn't it?"

"I don't know about that," Ellie replied, "There is a difference between God's word through the Bible, and God's word to you personally. The Bible has stood the test of time, multiple eyewitness testimony, consistency in its picture of God and Jesus, common use in the church, and historical fidelity. God's personal word to an individual has yet to meet even some of those criteria."

"Are you telling me that I should just ignore it then?" said Charles.

"Of course not!" retorted Ellie, "What did He say to you in the Bible? To you personally?"

"God told me that we are supposed to get married someday."

Thud! Ellie stomped off. Rubbing the bump on the back of his head Charles thought, "Boy, that ground is sure hard and cold."

What is this thing called Science?[5] This is the title of a popular introductory textbook on the philosophy of science. "Why should I be interested?" you might ask, "I am not a scientist." To be scientific carries considerable weight in our culture. To be in a field deemed "unscientific" or "scientifically

5. Chalmers, What is this thing called Science.

suspect" is to be insulted and put in a position to be ignored by others, whether merited or not. You believe that God created the world by some means. That makes you a creationist of some kind. But creationism is not scientific. Those who use any form of science in justifying their creationism are practicing a "pseudo-science," or are "unscientific" about the whole matter, which at this point, includes you. If I have correctly described you, how do you feel? Think you have been treated fairly? What is wrong with this whole train of thought? How would you argue against it? You may have dozens of questions and objections to this whole line of thought, but a good starting place would be to ask, "Just what do you mean by science and scientific? And why does being scientific about something make it inherently superior to other ways?"

The, *What is this thing called Science?* question gets even more challenging. In the early 1980s the state of Arkansas passed a law that that required public schools in the state of Arkansas to give as much equal time in discussing origin issues to creation science as they did to the standard evolutionary hypotheses currently in the textbooks. The resulting furor brought this law to court in Little Rock, Arkansas, in 1981. There the presiding Judge William R. Overton overruled the law by citing that creation science was unscientific. It was unscientific because it failed several essential characteristics of science: creation science was, 1) not guided by natural law, 2) cannot be explained by reference to natural law, 3) is not testable against the empirical world, 4) its conclusions are not tentative, and 5) it isn't falsifiable. Ignoring possible problems with the law itself we ought to ask, by what right does a legally trained judge get to decide what science is and is not? How good of a job did the good judge do? What are the ramifications of his opinion? The judge had to make a decision, and he chose the definition of science as a way to get there. His attempt though impressive at first glance, was hopelessly naive. Had his definition been taken as a legal precedent for all cases involving a definition of science, it could have caused huge swaths of very legitimate programs in science to have been declared unscientific and possibly ground them to a complete halt for lack of funding and support. But few scientists would have stood for such action, even though the majority of them, when shown his opinion would have agreed heartily with him initially. What!? Am I saying the majority of scientists do not have a very good definition of science and the scientific method? What I am arguing is that *no one* has a very good definition of science and the scientific method and it is not likely that anybody is going to come up with one anytime soon; if ever. This is crazy. Can you practice a field as successful as science and not be able to come up with a good objective explanation of what it is? The answer is yes, you can. Furthermore I will argue that it is very good that we cannot come up with a good airtight explanation of the beast. Let me explain . . . .

The approach philosophers of science have taken in getting to this point is to set up a model of science and how it operates, and then they tear it apart until a new model is dreamed up. This has been going on for over a century and a half until, finally in 1962 when Thomas S. Kuhn published the pivotal text, *The Structure of Scientific Revolutions*[6], philosophers of science started to realize that maybe they were looking up a blind alley. Rather than try to define science and how to demarcate it (separate it) from other fields, it was better to just study how it operated in the past and present and make general observations from the historical record. Let us forget trying to define science and what separates it from other disciplines (is it really separable?) and just concentrate on how science has operated and is operating. This is called the demise of the demarcation problem and the rise of the historical approach. How did we get here?

> On June 24, 1833, the British Association for the Advancement of Science convened its third meeting. Eight hundred fifty-two paid-up members of the fledgling society had traveled to Cambridge from throughout England, from Scotland and Ireland, and even the Continent and America, to attend . . . The atmosphere was charged with barely suppressed excitement and anticipation as the audience watched one of the speakers take his place on the stage before them. It was William Whewell—a tall, robust man in his late thirties, renowned for the brawn of his muscles and the brilliance of his mind. At Cambridge he was a star: outspoken fellow of Trinity College, recently resigned as Professor of Mineralogy, the author of a number of Physics textbooks and a new provocative work on the relation between science and religion. In less than a decade he would surprise no one by being appointed Master of Trinity, the most powerful position at the university; some would say the most powerful position in the entire academic world . . . He discussed the current state of the sciences, singling out astronomy as the "queen of the Sciences." . . . It was a masterful performance, just as the organizers had expected in inviting Whewell to open the meeting. After respectful applause—not only for Whewell, but for their own good sense and good taste in coming together as they had—the audience grew silent.
>
> As the applause died down, one man rose imperiously. It was, the other members realized with some surprise, Samuel

6. Kuhn, The Structure of Scientific Revolutions.

Taylor Coleridge, the celebrated Romantic poet. Decades earlier, Coleridge had written a tract on scientific method. Although for the last thirty years he had rarely left his home in Highgate, near Hampstead, he had felt obliged to make the long journey back to his alma mater for the British Association meeting. It would be the last of such trips; he died within the year. His intervention in the meeting would have far-reaching consequences for those who practice science, even to the present day.

. . . beckoning back to the close—knit relation between science and philosophy that had existed since ancient times—"natural philosophers." Coleridge remarked acidly that the members of the association should no longer refer to themselves as natural philosophers. Men digging in fossil pits, or performing experiments with electrical apparatus, hardly fit the definition; they were not, as he might have said, "armchair philosophers" pondering the mysteries of the universe, but practical men, with dirty hands at that. Indeed, Coleridge persisted, as a "*real* metaphysician," he forbade them the use of this honorific.

The hall erupted in a tumultuous din, as the assembled group took offense at Coleridge's sharp insult. Then Whewell rose once again, and quieted the crowd. He courteously agreed with the "distinguished gentleman" that a satisfactory term with which to describe the members of the association was wanting. If "philosopher" is taken to be "too wide and lofty a term," then, Whewell suggested, "by analogy with artist, we may form *scientist*."

That the coining of this term occurred when, where, and by whom it did was no accident; rather it was the culmination of twenty years of work by four remarkable men, Whewell and three of his friends. It was also, in some ways, merely the beginning of their labors, for the term, thus launched, was not widely used for decades more.[7]

Science did not begin being called science. The term scientist was not given publically until 1833. Isaac Newton and other scientists of his era and earlier would have called themselves natural philosophers. They philosophized and studied the natural world. As the natural philosophers became accepted as a separate discipline from other academic endeavors, attempts were made to define what was unique about its apparent success. Natural philosophy as it was practiced in the ancient Greek and Roman world started with observations about the natural world and then general conclusions were drawn about the way things operate. From these general suppositions, logic was used to deduce specific conclusions about individual

7. Synder, The Philosophical Breakfast Club, 1–3.

specific things. The emphasis was on the logic and reason used to get from the general to the specific conclusion. We call this deductive reasoning. Celestial bodies moved in a perfect heavenly realm, and thus had to move in a perfect geometric fashion, i.e. a circle or series of circles, not an ellipse or other imperfect shape. Each substance sought its own kind, hence earthen objects sought the earth and its center and moved downward, and fire, being a light kind of substance, sought its kind and rose up in the direction of the heavenly spheres and light above us. As long as the general conclusions were correct and your deductive reasoning accurate, you could get pretty far with this method. But if the general conclusions were flawed and your deductive reasoning a little off, then the results could be pretty bad if you did not check and correct them by other means.

With the fall of Rome to outside invaders, much of this "ancient science" was lost. The newly Christianized Europe spent its first 500 years stabilizing its borders from outside raiders and creating a new culture and worldview. As a stable culture emerged, Europe started a university education system in 1150—1200 AD, based on a merger of the old Greek and Roman classics that survived the fall of the Roman Empire and the new study of Holy Scriptures. These were the universities at Oxford, England; Paris, France; and Bologna, Italy; and others throughout Europe which exists to this day. The science and scholarship that developed from these schools was at first an imitation of the ancient natural philosophy under the overarching philosophical guidance of Christian theology. It was called scholasticism by scholars, but it quickly developed a life of its own. Theological conflicts with the pagan assumptions of the old Greek and Roman philosophers caused medieval scholars to re-examine their disciplines and change accordingly. In the years before the Renaissance, emphasis was moved away from a deductive approach to the study of nature to a more empirical and inductive approach. With the Renaissance, this change became a torrent and modern science was born. By the seventeenth and eighteen centuries science, "natural philosophy," was successful enough to merit attempts to explain how it worked. A first formal attempt that still exists to today is called empiricism. The British empirists John Locke, George Berkley, and David Hume held that *all knowledge should be derived from ideas implanted in the mind by way of sense perception.* The *positivists* had a broader view of the facts but agreed that *knowledge should come from the facts of experience.* Between 1924—1936 in Vienna, Austria; a particularly influential group of positivists, called logical positivists, gathered around the philosopher Moritz Schlick, and were called the Vienna circle. They believed that beliefs must be justified on the basis of experience. Rudolph Carnap, of this group, made the most significant attempt to formalize this philosophical approach.

He set out a verification principle which accepts only statements that are capable of being verified. Religious statements about God or otherwise are dismissed as "meaningless" because they cannot be empirically verified.

Though fairly old, empiricism and positivism are still very common and current views about the nature of science. Breaking it down further, we could say science is based on the facts of experience and that,

1. Facts are directly given to careful, unprejudiced observers via the senses.

2. Facts are prior to and independent of theory.

3. Facts constitute a firm and reliable foundation for scientific knowledge.

Look at the diagram below.

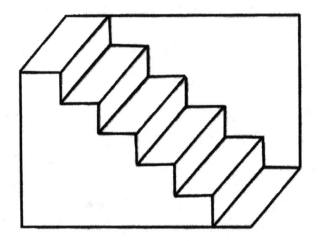

Are you looking down at the staircase? Or looking up as it passes over your head as in a closet underneath the stairs? The exact same lines exist in both staircases and possibly even on the retinas of two people observing this drawing. But they see two different images or "interpretations" of that image. To make matters worse, show this same drawing to a third-world tribesman who has never seen a drawing or a staircase, and they will not even *see* a staircase. The astronomer who discovered the planetoid Pluto was examining stellar telescopic photos in the region of space he expected to find the next unknown planet, and sure enough he discovered a star that moved, and found Pluto. A few months earlier another group of astronomers who were not looking for Pluto examined the same region of space over time, and did not discover Pluto. They didn't see it. Compare your first experience looking through a microscope with an experienced biology teacher. The teacher could easily see a specific part of the cell. You had great trouble "at first" seeing it. What we see is very conditioned by our experience, expectations, and prior training.

The Hungarian scientist and philosopher Michael Polanyi gives the following classic example of the relationship between our experience and what we see.

> Think of a medical student attending a course in the X-ray diagnosis of pulmonary diseases. He watches, in a darkened room, shadowy traces on a fluorescent screen placed against a patient's chest, and hears the radiologist commenting to his assistants, in technical language, on the significant features of these shadows. At first, the student is completely puzzled. For he can see in the X-ray picture of a chest only the shadows of the heart and ribs, with a few spidery blotches between them. The experts seem to be romancing about figments of their imagination; he can see nothing that they are talking about. Then, as he goes on listening for a few weeks, looking carefully at ever-new pictures of different cases, a tentative understanding will dawn on him; he will gradually forget about the ribs and begin to see the lungs. And eventually, if he perseveres intelligently, a rich panorama of significant details will be revealed to him; of physiological variations and pathological changes, of scars, of chronic infections and signs of acute disease. He has entered a new world. He still sees only a fraction of what the experts can see, but the pictures are definitely making sense now and so do most of the comments made on them.[8]

It is absurd to think that statements of facts enter our brain by way of observation. All observations must be interpreted somehow before they become a statement of fact.

Once upon a time there was a rich young man who wanted to leave his mark on society. He decided to dedicate his life to science. He would rise every morning, open his scientific notebook, and record every fact he could as he went through the day. The temperature changes, the wind speed and direction, amount of cloud cover, the number of people around him, and the things they said. Day after day, month after month, year after year, he recorded every fact he could and filled volume after volume of the scientific notebook with his observations. In his old age, with satisfaction, he gazed

---

8. Polanyi, Personal Knowledge, 101.

at room after room filled with his notebooks. He decided to donate it all to the National Academy of Science after he died. He died, and sure enough his estate executor approached the National Academy of Science about this treasure trove of scientific information. To the executor's surprise, the answer was, "Not interested." That was because the facts our fictitious rich man gathered were just a motley collection on any old fact, not relevant facts about a topic of interest. In fact most scientists spend a great deal of time "excluding facts" and influences so that their carefully designed experiments give just very specific facts.

What have we learned? Though we can observe common facts among different observers in science and other disciplines, this is not at all a given, and requires a great deal of training and practice to have some commonality in observed facts. Facts do not come into existence in a vacuum. We all have some theory of what those facts might mean before we even see them. Observations have to be interpreted before they mean anything and become facts, and that requires some kind of theoretical understanding before you even interpret the observation. So which came first? The theory or the observation? The chicken or the egg? So while science certainly uses observations and facts, and they are an integral part of it, defining science as based on the facts of experience is misleading. It grossly underestimates what is really happening. How many times have you heard a statement like this? *Science is about facts, religion is about values*, screams empiricism and positivism, doesn't it? Good science—bad science; observation—better observation; lousy theory—elegant theory; insightful publication—inaccurate publication. Scientists are making value judgments all the time and the basis of those value judgments are values that are impossible to justify by any empirical means. They are just flat accepted by the scientific community. If you go to a scientific conference and argue that faking data to support a theory is acceptable, if the theory really is right; you will get the same response as if you went into a church and argued that embezzlement is an acceptable means of supplementing the offering, as long as the money really is being used to help the poor. The response will be similar because the value systems violated in both cases will be similar, and equally unjustifiable by any simplistic empirical or positivistic definition of science.

A final thought on facts before we move on. Do facts constitute a firm and reliable foundation for scientific knowledge? In the centuries before Copernicus, it was a *fact* that the world did not move; a fairly obvious fact. It took centuries of hard thought and experimentation to understand the concept of inertia. It is possible for six billion people and a 10 mile deep layer of air to travel *with* the earth zipping through space at over 2000 miles per hour around the sun; and do it so that neither the people nor the air would be perturbed such that anyone could obviously tell the earth *was* moving. The *fact* of a stationary earth changed as our theoretical understanding of the solar system and physics changed. Sometimes scientific facts change as our theories change. It is hard to grasp this unless you study the history of science and see how many very commonly accepted scientific facts really have changed dramatically through the years. What evidence do we have that this is not still going on and will not continue to occur in the future?

So, maybe science is not just about facts. Perhaps it is about the process that scientists use to deal with their facts, fallible as facts are, that makes them different. Maybe scientific knowledge can be logically induced from careful and appropriate experiments. This requires that

- The number of observations is large,

- Observations are repeated under various conditions, and

- No accepted observation should conflict with theory.

Sounds reasonable. Let's look at the first one. What constitutes large? How many observations does it take to convince scientists of the destructive power of a hydrogen bomb? Just one it would seem. How many times would I need to stick my hand into the fire to convince others that it isn't very resistant to a blazing fire? How many measurements would it take to convince ornithologists that the average length of a mature male cardinal is 9.8 millimeters? Quite a few. It should be obvious that the number of observations or experiments it takes to be convincing depends heavily on what the hypothesis is, how the experiment is performed, and under what conditions. A scientist or community of scientists who decide that the induction is valid, have to use their personal judgment to evaluate whether a set quantity in "large" is really large enough. Note two key phrases in the previous sentence; community of scientists, and personal judgment. We will be coming back to those two phrases again and again, because they are both very crucial to the actual practice of science.

What about observations repeated under various conditions? What constitutes a significant variety of conditions? For example, do metals expand when heated? Do I test every possible metal, or just all types of metals? The answer is we draw on our current knowledge base to judge what constitutes significant. In short, scientists use their personal judgment. And when they try to publish the results, they have to rely on the community of scientists reviewing and judging their paper before the results become part of the common stock of human scientific knowledge.

Now examine the third point—no accepted observation should conflict with the theory. After Isaac Newton published his theory of universal gravitation it was discovered that the orbits of the moon and Mercury did not quite fit in with his theory. In the case of the moon, the problem was found to lie in an incorrect distance given for the distance between the moon and the earth. Once this was corrected, Newton's theory predicted the orbit of the moon quite nicely. However the orbit of Mercury was never made to exactly fit the equations of Newton's theory. For centuries this problem was known, but scientists universally praised Newton's theory and passed over this inconsistency. It was not until the early 1900s when Albert Einstein published his general theory of relativity that Mercury's orbit peculiarities were explained by referring to relativistic effects because of Mercury's close proximity to the sun. It is ridiculous to throw out a very successful theory just because a single observation negates it. The observation could easily be wrong, or some additional component of the experiment has not been properly accounted for that would make the experiment fit the theory. The personal judgment of the scientist, or the collective judgment of the reviewing community, must take everything into consideration before deciding whether a single observation or a series of them negates an entire successful theory. Every large and encompassing theory has problem observations kicking at the edges. Most will eventually be resolved, and persistent problems often pave the way to a new and better theory. To jump the gun by throwing out a theory is foolish. To ignore persistent problem experiments can stifle the advance of science. Here again, personal judgment arising from years of experience in the field will be the guiding factor in whether the theory is abandoned or not.

Some philosophers and scientists have argued that it is not the specific facts that makes science so unique and special, but the process by which scientists analyze these facts. Science uses induction to find correct and accurate answers. Is induction unique to science? Of course not. Induction has been used by philosophers, historians, and theologians for centuries. Actually science uses induction in a very limited fashion that merits a closer look. Scientists frequently use *modus tollens* logic. *Modus tollens* (shorthand for *modus tollendo tollens*) means in Latin, "the way that denies by denying." We can symbolize *modus tollens* logic as:

> H hypothesis
> I test implication of the hypothesis
>
> If H is true, then so is I
> But (as the evidence shows) I is not true
> H is not true
>
> *modus tollens* logic

Here is an easy example.

> H—All dogs have four legs
> I—Fido is a dog and has three legs
> Not all dogs have four legs

But

> If H is true, then so is I.
> (As the evidence shows) I is true
> H is true

This is not correct at all. It is an example of *the fallacy of affirming the consequent*. Here is a simple example.

H—All dogs have three legs	(is this true?)
I—Fido is my dog and has three legs	(true)
Thus all dogs have three legs	(true?? not necessarily!)

The fallacy of affirming the consequent.

Many people know this instinctively. Just because an experiment testing a hypothesis comes out showing that a proposed hypothesis is false doesn't at all mean that if the experiment came out with an opposite result the hypothesis would be proved true. It just means that the hypothesis survived being negated in this experiment and not necessarily for all future experiments. In a sense *modus tollens* logic is like a big box we run our conjectures through. The box tells us our conjectures do not work, at least under the experimental conditions given. It will not tell us that the conjecture is true. We have to infer that conclusion by running our conjecture through as many experiments and conditions as we deem acceptable and hope our conjecture survives it all. But if it does, then most likely it is correct, at least as we have *seen* it, *so far*. We are building up confidence that our conjecture, our hypothesis, is a correct representation of reality.

Here is another issue.

1- All cows have five legs

2- Rosie is a cow

3- Rosie has five legs

Perfectly valid deduction, *but* 1 and 3 are obviously false, because my initial premise is false. Allen Chalmers in his classic textbook on the philosophy of science summed it up as follows:

There is a strong sense, then, in which logic alone is not a source of new truths. The truths of the factual statements that constitute the premises of arguments cannot be established by appeal to logic. Logic can simply reveal what follows from, or what in a sense is already contained in, the statements we already have to hand.[9]

David Hume the famous early empiricist noticed another problem with inductive logic.

Principle of induction works on occasion X1
Principle of induction works on occasion X2
Principle of induction works on occasion X3
Principle of induction works on occasion X4
Thus the principle of induction always works

Obviously not so. Just because induction has worked 4 or 100 times does not guarantee that it will work the next time. Just because induction has worked pretty well where we have applied it so far, is no guarantee it will continue to work. How do we get out of this? Is all of science flawed to the core? Are our rockets to the moon just pure luck? My next tetanus inoculation will be fatal? No, it just means there is no such thing as a scientific proof, generally speaking. *Science just gives us statements of increasing or decreasing reliability.* Some scientific statements are very reliable and you are a fool to ignore them. Some statements are less so. The trick is finding out which ones are very reliable and which ones are not so reliable. Most of us rely on authority to figure that out: textbooks, academies of science, teachers, government, institutes, schools, universities, church, parents, etc. But that brings us back to two familiar terms; personal judgment and the community of science. And that injects a great deal of humanity into the entire process.

"Hello Ellie, how are you doing?" Ellie looked up at the inquiring face of Jon who asked the question. He had a cute face and personality but his leftist tendencies were known to all. How anyone could be an admirer anything that smacked of communism in this age after the fall of the Eastern bloc and the Soviet Union was beyond her. Maybe Jon did it just to get everybody's attention and be an iconoclast.

"I am doing fine, and you?" she intoned.

"Reading the Bible again Ellie?" he asked.

"Yes, you don't approve of it?" she replied back.

9. Chalmers, What is this thing called Science?, 43.

"Well it is not for me to approve of it or not, it's great literature, good morals and all that. I just don't see why you take it so personally." Jon stated. "I mean science has never found God. No crucial experiment that showed that He does exist, you know."

Ellie stiffened a bit and replied, "Like you are going to find a crucial experiment that demonstrates the existence of a being that created all things and stands outside the whole system. How can that which was created ever show or disprove something outside of itself? Remember Godel's incompleteness theorem; you can't prove the completeness or sufficiency of a system from within that system."

Jon stepped back a little, and said, "Hey don't fire both cannons at me! I never said you could prove or disprove God, I just wondered how you could take it so seriously when so many in the National Academy of Science don't." Now it was Ellie's turn to lean back in her chair a little.

"What do you mean?" she asked.

"Well in two studies, one at the turn of this century and another one a few decades ago showed that both times about 93 percent of the members of the National Academy of Science were either agnostics or atheistic in their personal belief about God" he replied.

"Oh really," she replied, "What about the other 7 percent?"

"Well they are theists it seems," he said.

Getting an idea Ellie asked, "Well what about the scientists not in the academy, how do they fare, how do they believe compared with the typical U.S. population"

Jon was quick to answer, "Well they pretty much match up with the rest of the population, about 7 percent agnostics or atheists, the rest theists, but remember 93 percent in the academy is a pretty overwhelming majority."

Ellie paused and then replied, "But all of the scientists in the country and the whole population are a pretty big majority compared to a few hundred in the National Academy of Science. The National Academy elects fellows to itself, don't they?"

"Well I guess it is a matter of who and what you want to take as your authority." Jon stated just a trifle harshly.

"I guess so." said Ellie.

Jon quickly said, "Remember, you cannot falsify God. You can't come up with an experiment whose results are such that you conclusively disprove the existence of God. You can always come up with a reason of some sort to explain the results away. Sort of like fairies and pop psychology, no matter how the results turn out, you can always come up with a reason on why your theory still fits. Einstein's theory of relativity made specific predictions. If the experimental results showed them to be wrong then his theory was wrong and he would have to go back to the drawing board. Now that is real truth; something that can be falsified."

Ellie thought a moment and said, "Well communism falls into the same category as fairies and pop psychology. You can interpret any historical or social event in history as a class struggle between the bourgeoisie and proletariat."

"Well they are still working on that." Jon replied. "Hey, I've got to get to class; we'll talk about this later." His voice dropped in volume a bit as he drifted down the sidewalk.

Ellie thought to herself, "Is he just teasing me, or does he really believe, like that?"

Karl Popper was educated in the 1920s in Vienna when logical positivism was in vogue among philosophers. He became disenchanted when he saw what he regarded as pseudo-sciences, Freudism and Marxism, try to justify their scientific nature by the same philosophical methods he and others had used to rationalize science's special place. The publication of Einstein's theory of special and general relativity with its easily falsifiable predictions impressed him greatly. Perhaps the ability to be falsified was the key to science's great success.

Science advances by falsifying theories. Conjectures and theories are advanced as wished and falsification is the criteria by which theories are falsified. Falsificationists freely admit observation is guided by and depends on theory. A theory can never be said to be true, just the best available so far and only the fittest theories survive.

So what exactly is meant by a falsifiable statement or theory? For example, here are some easily falsifiable statements. *It always rains on Fridays.* Just wait and watch the weather on a series of Fridays, and you will have falsified the statement. *All solids expand when heated.* Find one solid that does not (remember those heat shrinkable plastics) and this statement is falsified. *All masses attract each other by gravity.* This is easily falsified but so far no one has done it. Some masses repel each other (electrostatically charged opposite ones for example), but the repulsion has never been traced to a lack of attraction due to gravity. So this is a falsifiable statement that has resisted falsification for centuries. That would make it a very reliable scientific statement. For a theory to be falsifiable then it would need to make a lot of falsifiable statements and predictions that can be tested. If those statements and predictions resist being falsified then the theory has not been falsified. As more and more predictions and statements become falsified, then the theory becomes more and more falsified.

ability to test or fail

What are some non-falsifiable statements? *It will snow tomorrow or it will not.* That one is pretty obvious, how can that statement not be true? *All points on a Euclidean sphere are equidistant from the center.* Well if we squish the points of a Euclidean sphere into the shape of a rectangle, the points won't all be equidistant from the center. But then it is no longer a Euclidean sphere. In fact by definition, the points have to all be equidistant from the center or it isn't a Euclidean sphere. The statement is self-fulfilling and non-falsifiable, and thus a non-meaningful statement. So when someone says the concept of God or the idea of religion is non-falsifiable; they are making quite a slam against God and religion.

When Karl Popper formulated his ideas about the essence of science, he suggested that falsification of theories become the important landmarks and points of achievement in understanding science. Theories that are easily falsifiable should be preferred over those which are not. Science has been a continual process of advancing theories, falsifying them, and then replacing them with better theories that may someday themselves be falsified and replaced. But it is the process and method of falsification coupled with the advancing of new ideas and theories that makes science so unique and compelling as a way to truth. The idea has a certain amount of historical appeal.

How many times have we read of the crucial experiment that forever killed the credibility of an older theory, and established a new one? Almost every sophomore modern physics textbook tells of the Michelson and Morley experiment. They used a large interferometer to test for the presence of the universal ether medium that was supposed to support the propagation of light and other electromagnetic waves. Water waves require the medium of water to support them. Sound waves have to have air before they can propagate. Thus electromagnetic waves must have some kind of medium in which they propagate, it was initially thought. If the interferometer beams were swung such that one beam was perpendicular to the direction of travel of the earth and measured the velocity of light, and then moved it into the line of travel of the earth and re-measured the velocity of light, they should see a sizable difference in the velocity of light that they measured. This was because the earth was traveling through the stationary "electromagnetic ether," it was thought. No sizeable difference was measured; the theory of a universal ether that light moved in was falsified and abandoned; and Einstein developed the special theory of relativity to explain the negative result of the Michelson and Morley experiment. As a side feature, the highly successful aspects of Newton's theory of gravity could be shown to be a subset of results one obtains from relativity theory if we stay at very low speeds. So the new theory explained everything that Isaac Newton's theory did, plus additional things that Newton's did not. At least this is the way the progress of science is presented in many textbooks and lectures. What really happened?

Actually Einstein had antagonized his professors at his Ph.D. institution, the University of Zurich in Switzerland, and was not awarded a university job after his doctorate, as was the custom at that time. He found an easy job in the patent office in Switzerland and in his spare time studied physics articles. A review article of that era discussed the major problem areas of physics. In two short years in his spare time, Einstein wrote several articles that addressed and often *solved* those problem areas. Within a decade, his name was a household word and he held a prestigious university appointment. One of the problem areas was how to resolve the inconsistencies between the new field of electrodynamics and Newton's theory of mechanics. Electrodynamics studied radio waves and how light, electricity, and magnetism are related. Mechanics studies how masses are attracted to each other by gravity and thus how they move relative to each other. Throw a rock; what does the trajectory look like? Einstein realized that if you could step aboard a rocket-ship traveling the speed of light and look out the window at a light beam from a flashlight beam someone on the ground shined up at you after you took off, you would see a very strange sight. Light is an electrical field and magnetic field interacting with each other to produce what is called an electromagnetic wave that travels at the speed of light. Einstein realized that if you and the wave were traveling the same speed you would just see (assuming you could see electrical and magnetic fields) an electrical or magnetic wave standing perfectly still, and varying in magnitude. This was ridiculous according to electrodynamic theory and so Einstein knew something was drastically wrong. Either Newtonian mechanics or Maxwell's electrodynamic theory was irrational and Einstein decided (personal judgment) that Newton was badly missing at this point. He decided to fix Newtonian mechanics and relativity was born. In fact Einstein was not even aware of the Michelson-Morley experiment until his paper was ready for publication. A reviewer pointed it out, so Einstein included it in his list of references. At this point you are wondering where is the crucial experiment or series of experiments that led Einstein to his conclusions. Is the thought experiment the crucial one? Was 250 years of Newtonian mechanics falsified as the result of a thought experiment and a belief that Maxwell's electrodynamics was inherently more rational and stable than Newtonian mechanics? If so, Karl Popper's theory of scientific advance by falsification has some serious problems.

Falsification has other serious problems. Falsificationists say observation and theory are linked. You need a theory to interpret your observation and observations support or falsify your theory. So if an observation is true, then your theory is false. Then you find your observation to be true. But if observation somehow depends on theory, how can you conclusively say the theory is false? The theory you are trying to falsify is somehow justifying your observation. If it is false then how can you believe your observation?

In short, conclusive falsification of theories by observation is not possible. You never completely falsify a theory. There is always an element of doubt in every experiment you do. A theory cannot be conclusively falsified because the possibility always exists that some part of the complex test situation, *other than the theory under test*, is responsible for the falsifying result of the experiment. This is called the *neglect of auxiliary hypotheses problem*. This experimental issue was first raised by the French physicist Pierre Duhem at the start of the twentieth century. It was re-introduced into philosophical discussions by the Harvard philosopher Willard Quine and is called the Duhem-Quine thesis. An example may help illustrate the issue. Below is a diagram based on a similar diagram in Thomas Kuhn's famous book *The Structure of Scientific Revolutions*. It shows the earth in orbit around the sun with the North and South Pole tipped at an angle to the earth's orbit. This angle is what gives us our seasons during the year. If we look at a star in the northern sky at night relative to the direction of the earth's North Pole you will see a certain angle given between the two. If that season is winter, and then you wait until summer, the angle should change a little as shown in the diagram. The difference between those two measurements is given by the angle P in the diagram and is called the parallax angle. In other words, between summer and winter we should see a little "wiggle" in their positions relative to each other. To see this more clearly, hold your hand out at full length from you with your thumb pointing up. Then have a friend stand about twenty feet from you. Close your left eye and place your thumb over your friend. Now close your right eye and then open your left eye.

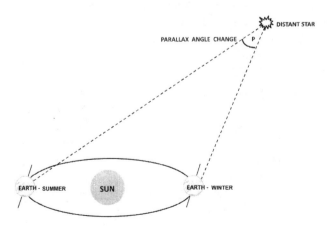

Is your thumb still on top of your friend? No, it will have jumped to the side because of the different sighting angle. Now have your friend move one hundred feet back and repeat the experiment. Then have them move 1000 feet back and try it again. Each time as they move further back you will see the amount of jump become smaller and smaller until eventually if they move far enough back you cannot detect the jump or parallax error. If we move the stars far enough away from the earth, you cannot detect the star's parallax (or stellar parallax) from summer to winter with the naked eye. When Copernicus proposed his heliocentric theory of the solar system, the best observational astronomer of the day Tycho Brahe reasoned that Copernicus' theory should give seasonal stellar parallaxes. Using the best astronomical equipment of his day Tycho looked for these stellar parallaxes and found none. As a result he declared Copernicus' theory to be wrong and formulated his own theory of the solar system which had the planets circling around the sun, but the sun, moon, and stellar sphere circled around the earth which was at the center of the solar system. A perfectly reasonable scientific conclusion, but unfortunately it was very wrong. His theory lies dormant in the dustbin of failed theories. What went wrong? The medieval astronomers knew the stars were far away from the earth. The medieval taught that if you could walk up into the sky and walked 40 miles every day for eight thousand years, you still would not reach the stars[10]. But the idea that the stars were so far away that it would take billions and billions of years to walk to them was beyond their wildest ideas, and beyond Tycho Brahe's too. So when Tycho designed his stellar parallax experiment, he had *built into his experiment the auxiliary hypothesis* that the stars were only thousands of walking years away, not billions and billions of years. More than likely it was such a part of his worldview that he did not even realize that he had built that particular auxiliary hypothesis into his observations. Not only was he faced with this kind of worldview blindness, but all of his astronomical peers were also. It was decades later that Galileo helped firmly establish the reliability of the heliocentric theory with his telescopic observations. Then the astronomers of the day were forced to realize that the lack of observable stellar parallax meant the stars had to be further away than ever dreamed. It turns out you can see stellar parallax; but it wasn't until 1838 when telescopes were powerful enough to detect them. Poor Tycho couldn't use telescopes; they weren't invented yet. Can we blame him or sneer at his work? Would we have done any better had we been in his situation? Most likely not.

10. Lewis, Discarded Image, 98, Lewis quoting the South English Legendary, Ed., C. d'Evelyn, A. J. Mill (E.E.T.S., 1956), vol II, 418.

Now here is the scary part of this historical saga. Are we really in that much a different position today? We have a lot more knowledge, much better technology, much better training and understanding. But compared to the future, will we look like Tycho Brahe does to us today? Will they be as sympathetic to our science? What auxiliary hypotheses have we accidentally built into our experiments that have led us far astray from the truth; that we *hope* will be revealed in the future. To be honest, there is no such thing as an absolute scientific proof. Every experiment we do has hanging in the background the nagging fear: have I forgotten something that is crucial to this experiment? All we can do is to do our best, and assume we have done a good job and go from there. What else can you do? We have *faith in our personal judgment* and vision of science that we have done a good job, and that our results are meaningful and useful. Now this is a strange result; we have *faith* in our science. What are faith and science doing couched together in the same sentence?

Karl Popper admitted that it was often necessary for scientists to retain theories in spite of apparent falsifications. Copernicus did, and with an almost religious fervor. So though ruthless criticism and attempts at falsification of theories is recommended, a certain amount of dogmatism and confidence in one's theories has a very positive role to play in the development of science. Well, if this is the case one might wonder what is left of falsification once any significant amount of scientific dogmatism is allowed a key role in the development of successful theories. A well-known twentieth century scientist/philosopher named Michael Polanyi has much to say on this point of dogmatism in science that we will pick up on later.

It is obvious that this line of thought goes further than scientific theories. If science cannot be supported by a falsification approach, then on what grounds do we reject God and religion because they at first glance appear to not be falsifiable? A common approach taken by some non-theists is:

1- God is omnipotent (all-powerful) and omniscient (all-knowing)

2- God is completely good

3- The world contains instances of suffering and evil

4- (Then) A good omnipotent God would eliminate suffering and evil (He has not; so God must not exist)

This is a very common objection to God among many intellectuals. If we apply the Duhem-Quine thesis then it becomes apparent that this, like many "experiments," is not provable at all. Because you cannot assume that statements 1, 2, and 3 are all that is involved; particularly with a being like God who is completely above anything in the universe. What auxiliary hypotheses have we neglected in playing thought experiments like this with God? Alister McGrath has a very readable discussion of this interplay of logic and religion in his *Science and Religion*[11] text.

It is becoming very apparent that theory, observation, facts, human preconditioning, human experience, our worldview, and how we pursue and define science are very interrelated. It is very hard to separate our humanness from our science. But why? Michael Polanyi expresses the idea beautifully in his magnum opus (Latin definition "a great work"), *Personal Knowledge*.

> In a full 'main feature' film, recapitulating faithfully the complete history of the universe, the rise of human beings from the first beginnings of man to the achievements of the twentieth century would flash by in a single second. Alternatively, if we decided to examine the universe objectively in the sense of paying equal attention to portions of equal mass, this would result in a life-long preoccupation with interstellar dust, relieved only at brief intervals by a survey of incandescent masses of hydrogen—not in a thousand million lifetimes would the turn come to give man even a second's notice. It goes without saying that no one—scientists included—looks at the universe this way, whatever lip-service is given to 'objectivity'. Nor should this surprise us. For, as human beings, we must inevitably see the universe from a centre lying within ourselves and speak about it in terms of a human language shaped by the exigencies of human intercourse. Any attempt rigorously to eliminate our human perspective from our picture of the world must lead to absurdity.[12]

These words were penned in 1958, four years before Thomas Kuhn published his landmark, *The Structure of Scientific Revolutions* in 1962. Kuhn's book was a major turning point in the philosophy of science debate. The philosophers and *some* scientists started realizing that pursuing an exact definition of science was futile, and that it may be better to try to understand how science has operated and functioned at different times in the past and present.

Thomas Kuhn was a theoretical physicist who switched from physics to the history of science. As he studied the history of science, particularly during turbulent periods, he became convinced that a distinct pattern of scientific advance was displayed historically. He published his findings in *The Structure of Scientific Revolutions*. It was in this text that the term paradigm

---

11. McGrath, Science & Religion, 120–27.
12. Polyani, Personal Knowledge, 3.

shift was first discussed. The book became very influential and started a major shift in scientific philosophical thinking from trying to define science from its methodology to trying to define it from an understanding of the theoretical frameworks in which scientific activity takes place, *i.e. its history*. What started Kuhn's thinking was partially initiated by his historical discovery that many out-of-date scientific beliefs that are now called superstitions and myths were produced by the same methods that now lead to scientific knowledge. He mentions Aristotelian dynamics, phlogistic chemistry, caloric thermodynamics as examples, and we could add horoscope casting, and divine creation of individual species as reasonable additional examples. Here is Kuhn's quotation.

> If these out of date beliefs are to be called myths, then myths can be produced by the same sorts of methods and held for the same sorts of reasons that now lead to scientific knowledge. If, on the other hand, they are to be called science, then science has included bodies of belief quite incompatible with the ones we hold today[13]

Kuhn's picture of how science proceeds is as follows:

prescience—normal science—crisis—revolution—
new normal science—new crisis—

Pre-science is distinguished from normal science by severe disagreement over the fundamentals. There are almost as many theories as there are workers. Each theoretician has to justify his own approach. It is almost impossible to get down to normal work, fleshing out the details of the theory. However as the field matures, there emerges a normal science with a fundamental paradigm and a set of teaching tools to propagate the paradigm.

Normal science as Kuhn views it is research firmly based on one or more past scientific achievements. These achievements are what the scientific community acknowledges as supplying the foundation for its further practice. Some examples past and present are Aristotle's physics, Ptolemy's astronomy, Newtonian dynamics, Lavoisier's chemistry, quantum mechanics, and plate tectonics. Some classic textbooks expounding Kuhn's normal science are Aristotle's *Physica*, Ptolemy's *Almagest*, Newton's *Principia* and *Optics*, Franklin's *Electricity*, and Lyell's *Geology*. Normal science uses a paradigm, an accepted model, theory or pattern. The paradigm is an object that needs further expression and specification under new conditions. It is a promise of successful discovery in selected future examples. Often the paradigms and laws are so interwoven that natural laws can be predicted even

---

13. Kuhn, The Structure of Scientific Revolutions, 2.

before further experiment. Kuhn claimed that normal science had three types of problems: determination of significant facts (about the paradigm), matching the facts with the theory, and expression and extension of the theory or paradigm. After the initial discovery and acceptance of the new paradigm, these three types of research are basically considered mopping up operations. Mopping up operations is what engages most scientists through their careers. The community of normal science and those scientists who practice it are often intolerant of new theories invented by others.

As an example; a molecular spectroscopist (like the author) has two fundamental paradigms that are used to understand molecular spectra. One is classical Newtonian mechanics and the other is quantum mechanics. Using the laws of Isaac Newton, they can pretend a molecule is like a very tiny set of point-like masses that are connected together with tiny coiled springs. The ball masses represent the atoms and the coiled springs represent chemical bonds that hold them together. If the molecule is shaken, the whole assembly of balls and springs will vibrate back and forth, and shake around in a specific way that can be determined by the laws of Newton, and the properties of coiled springs. However, real molecules absorb light energy and can be "excited" to higher levels of energy (that is—they vibrate and bounce around a whole lot more). But they do it in a very specific fashion that is not governed by Newton's laws but by another theory of motion and dynamics that works at very, very tiny levels, i.e. atomic sizes. This theory was discovered in 1926—1928 and is called quantum mechanics. It claims that energy and motion at the very tiniest levels is packaged in discrete bundles or quanta and so changes in energy levels must be in discrete jumps or levels. Most of my scientific career has been spent in examining new and different molecules and showing how the current theory can explain a particular molecule's spectrum just like it has other molecules. I just keep moving to newer and more complicated molecules; molecules that could never be examined with past technology, and molecules that are interesting for their medical and biological properties. But the basic paradigms I use have not changed much at all. This is normal science, fleshing out the standard paradigms. If I had to review a journal manuscript that explained a molecular spectrum with a new paradigm, I would be very suspicious and likely not pass the manuscript unless I had book upon book, and scientific news releases verifying that the new paradigm really did work. I would have to read them for myself before I would believe the new paradigm.

The crisis stage is reached when anomalies build up that strike at the very heart of the old paradigm; they resist all attempts to remove them. A professional crisis can set in that while very disturbing, can set the stage for the new paradigm. Some examples are Galileo's telescope observations of the moon that showed "flaws" on what should have been "perfect" heavenly body; and the moons of Jupiter that showed the earth was not the only center of revolution in the universe. A classic crisis is often portrayed in the development of quantum mechanics during 1926–28. The laws of Newton, when applied to the electrons and protons in the atom, failed miserably, and a whole new type of "mechanics" called quantum mechanics was developed to handle atomic phenomena. The insecurity of scientists before the solution of this scientific crisis is fascinating. Below is a quotation from the Nobel laureate Wolfgang Pauli in the years before 1926.

> At the moment physics is again terribly confused. In any case it is too difficult for me, and I wish I had been a movie comedian or something of the sort and had never heard of physics[14]

Five months later after Werner Heisenberg published his wave mechanics paper that laid the foundations for quantum mechanics, Pauli quoted.

> Heisenberg's type of mechanics has again given me hope and joy in life. To be sure it does not supply the solution to the riddle, but I believe it is again possible to march forward.[15]

Crisis in science is resolved in three ways. The old normal science solves the anomalies problems and marches on as before. No solution is found and the problem is postponed for future generations to solve. A new paradigm emerges and a battle ensues over its acceptance. The change of scientists' allegiance from one paradigm to another was labeled by Kuhn as a Gestalt switch (from Gestalt psychology) or "religious" conversion. Other authors have called this a *paradigm shift* and applied it to areas beyond science as well. Eventually if the new paradigm succeeds, it takes over, and becomes the new normal science with its own set of textbooks and teaching examples, etc. Max Planck, a Nobel laureate in physics, offered another explanation of how the new paradigm succeeds.

> A new scientific truth does not triumph by convincing its opponents and making them see the light, but rather because its opponents eventually die, and a new generation grows up that is familiar with it[16]

This is a more human and perhaps more cynical view of how science can advance, but coming from a scientific giant like Max Planck, it merits

14. Kuhn, The Structure of Scientific Revolutions, 84.

15. Kuhn, The Structure of Scientific Revolutions, 84.

16. Plank, Scientific Autobiography and other Papers, 33–34.

some serious reflection. But what makes a scientist embrace a new paradigm in the early stages when it is not at all apparent that it will succeed? If
it does, the scientist becomes a hero to future generations, and a scientific
father to the new paradigm. If it fails, then his scientific life is wasted and
his scientific reputation can lie in ruins. It is not pure facts that make a man
jump to a new paradigm this early, but something not objective or articulate
that drives him. Kuhn wrote:

> The man who embraces a new paradigm at an early stage must
> often do in defiance of the evidence provided by the problem-
> solving. He must, that is, have faith the new paradigm will suc-
> ceed with the many large problems that confront it, knowing
> only that the older paradigm has failed with a few. A decision
> of that kind can only be made on faith . . . This is not to suggest
> that new paradigms triumph ultimately through some mystical
> aesthetic. On the contrary, very few men desert a tradition for
> these reasons alone. Often those who do turn out to have been
> misled.[17]

Here again we see another unusual link between faith and science.
Kuhn argued that philosophers were accustomed to viewing science as
slowly working its way to some goal. But he argued that there really did not
have to be any such goal. Science could just be working from one paradigm
to another with no particular end in sight other than a new state of knowledge of the community. Scholars quickly charged Kuhn with presenting a
relativistic account of science and its progress. Kuhn distanced himself from
these charges in postscripts in later editions of the book. But he never really
answered how we can decide that one paradigm is better than another. What
set of standards do we appeal to decide which paradigm is better? In what
sense can science be said to progress through revolutions? This philosophical hole just gets worse with later philosophers of science. But can we really
say science is just meandering around according to the dictates and cultural
values of a given culture and society? Is Roman surgery and medicine just
as "good" as modern medicine? Is Greek astronomy better than modern
astronomy that put a man on the moon and discovered new planets? Why
do we have this innate value sense that we really have gotten closer to a real
ultimate objective truth and we really are making progress? Is there a real
truth and a real set of values and ethics out there that really does guide our
frailties in the direction of ultimate truth?

Imre Lakatos suggested that not all parts of science are so equal. Some
laws are more basic and important than others. Lakatos built on Kuhn's

17. Kuhn, The Structure of Scientific Revolutions, 158.

work using a historical model that he hoped would eliminate the relativistic aspect of Kuhn's theory. Lakatos proposed that science is best understood as research programs with hard core assumptions and a protective belt. These programs had a positive heuristic that specified what a scientist can do and a negative heuristic that told scientists what they are not advised to do. The specification of a research program that indicated its merit or success was its ability to lead to novel predictions that are confirmed, and to offer a program of research. Progressive research programs retain coherence and occasionally lead to confirmed novel predictions. Degenerating research programs lose coherence and fail to lead to new confirmed predictions. The replacement of a degenerating research program for a progressive research program is Lakatos's version of a scientific revolution, or paradigm shift. With Kuhn later paradigms were deemed better than earlier ones because the scientific community judges them to be so. "There is no assent higher than the relevant community," Kuhn quoted. This is very relativistic and Kuhn was not happy with it, and Lakatos was very dissatisfied. So Lakatos sought a standard that was outside of the paradigms that could be used to identify in a non-relativistic way how science progresses. Progress was re-placing *degenerating research programs* with *progressive research programs* where progress is defined as *better predictors of novel confirmed phenomena*. Lakatos, like Kuhn, believed any philosophy of science had to adequately explain the history of science.

What is the hard core of Lakatos's research program? A study of the history of science shows what it is by the accepted tinkering with the hard core assumptions. There were serious attempts to modify Newton's inverse square law (basic mathematical law of gravitation) to make the orbit of Mercury fit. Copernicus did not place the sun at the exact center of the circular orbits in order to fit the observed data better.

Lakatos recognized that some scientists stick to degenerating research programs and often bring them back to life. Copernicus's sun-centered model of the solar system degenerated for 100 years because of its failure to fit the observed apparent size change of the planets as they moved toward and away from earth, and the lack of stellar parallax. Galileo and Kepler brought it dramatically back to life. This illustrated a serious problem for Lakatos's attempt to find an indicator of scientific progress. If a degenerating research program *can* be brought back to life, then when do we decide that it is *progress* to abandon a particular research program for another one? Lakatos suggested that Isaac Newton changed scientific standards in way that is progressive, better. *But with respect to what standard was the change progressive?*

As modern academia slipped into very relativistic post-modern thinking, some philosophers of science applied this thinking to science. Paul

Feyerabend was an Austrian at the University of Berkeley. He spent part of his career interacting with (and antagonizing) Karl Popper and Imre Lakatos. He published *Against Method: Outline of an Anarchistic Theory of Knowledge in* 1975.[18] Feyerabend tried to undermine all characterizations of method and progress in science. In his analysis of Galileo's book *Dialogue*, where Galileo defends the heliocentric theory of the solar system, Feyerabend argues that Galileo had to resort to rhetorical tricks and propaganda because his facts were so weak, and the reliability of his telescope observations so bad. Since philosophers of science had failed to define a universal scientific method, he argued.

> What remains are aesthetic judgments, judgments of taste, metaphysical prejudices, religious desires, in short what remains are our subjective wishes.

In short; there is no scientific method. Scientists should follow their subjective wishes. Anything goes. Feyerabend's theory is couched inside the ethical framework of John Stuart Mills of the nineteenth century. Mills placed such a high value on individual human freedom that no severe constraints are placed on the individual's ability to choose what he wants and how he wishes to achieve it. Feyerabend called Lakatos a fellow anarchist in disguise and said Kuhn's appeal to scientific consensus was wrong because he could not distinguish between legitimate and illegitimate ways of achieving consensus (i.e. kill off all your scientific opponents).

The problems with Feyerabend's approach are legion, but it does illustrate where thinking can go when you decide that the *only way* to find truth is through human reason. What are some of these problems? Feyerabend's notion of freedom is entirely negative; freedom is understood as freedom from constraints. But members of a society must give up some privileges in order to gain access to other privileges. The world is not one vast container of privileges waiting to be picked at random, with infinite resources available to fulfill them for everybody. Also, by what standard does Feyerabend apply legitimate and illegitimate to the various ways one could achieve consensus (i.e. kill off all your scientific opponents)? What higher standard do we use to determine the legitimaticity of a particular scientific consensus? Why do we always find ourselves appealing in some form to a higher standard? Maybe because it really is there; and it is impossible to have a genuine discussion without one. Even Feyerabend when he claimed that Kuhn could not distinguish between illegitimate and legitimate ways of finding consensus was *implicitly* appealing to such a standard by even

18. Feyerabend, Against Method.

using the words legitimate and illegitimate. Without some kind of standard of good or bad, right or wrong; the words legitimate and illegitimate would have no meaning. C. S. Lewis in *Mere Christianity*[19] says that there is a Law of Human Nature that exists and is apparent to all humans. It is the only law that we humans can choose to disobey, i.e. the law of right and wrong. We instinctively know that it exists when we claim we have been wronged by someone else. Without this universal law of human nature, the concept of being wronged would have no meaning. Thus a universal standard does exist, and Feyerabend is implicitly appealing to it.

Time for elaboration. Many people instinctively recoil at the thought of not being able to find this universal standard from purely rational methods. Yet we instinctively use a standard to recognize that modern science has progressed and gotten better, certain things are more beautiful, true, and right; we do compare and contrast things in our lives relative to this "standard" all the time. Why is simplicity and beauty in a theory preferred? In fact there is an amazing uniformity in all major civilizations with regard to certain fundamental assumptions about what is good or bad.

Imagine a world of 500 years ago where the scientific method was determined according to Feyerabend's criteria. Where would we be now? Would man have visited the moon? Would we be dead of smallpox? Who in their right mind would want to live in such a civilization? Alan Chambers, the author of *What is this thing called Science?*, rejected Feyerabend's anarchism and extreme relativism and then made this fascinating statement.

> I reaffirm that there is no general account of science and scientific method to be had that applies to all sciences at all historical stages in their development. Certainly philosophy does not have the resources to provide such an account.[20]

This brings us back to where we started; when I first asserted that *no one* has a very good definition of science and the scientific method. Chambers stated that if he was going to defend a change in the scientific method that avoids extreme relativism then he had better say in what way such a change is for the better. He said that unless there are super-standards for judging changes in standards then those changes cannot be taken in a non-relativist way. Summarizing, some philosophers of science argue that either we have universal standards or relativism; there is no middle ground. I agree and come down strongly on the universal standards or super-standard side of the issue. The incidental evidence for it is very compelling and my Christian

19. Lewis, Mere Christianity, 17–35.
20. Chalmers, What is this thing called Science?, 247.

beliefs make it obvious. Saying that philosophy cannot provide a general account of the scientific method does not mean it does not exist, just that we cannot describe it at this time. What is the problem with saying that there are universal standards by which we can judge and evaluate things? That we may have trouble articulating them does not mean they do not exist. The fact that many scholars find relativism repulsive may say something about an implicit admission that universal standards do exist.

Many, if not all, great academic questions eventually boil down to this. Is there something other than ourselves we can appeal to for higher standards by which we evaluate the world, or are we totally on our own? How you answer that question will determine the stream of thought one uses to evaluate our knowledge and world. Examine the picture below.

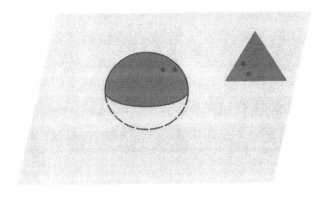

How would a two dimensional triangular creature evaluate a three dimensional spherical creature, that invaded his two dimensional world? If the two dimensional creature could just see in two dimensions, it would be impossible to fully see and understand the three dimensional creature as it is. This is a simplistic illustration of the difficulties we have in fully understanding what science is and how it works. If science's ultimate basis is outside of ourselves rooted in the fullness of God, why should we expect it to be easily definable and understood? Science is after all a very human activity that had its modern basis rooted in a heavily Christianized civilization—the Middle Ages.

3

The Society of Science

"WELL, HELLO JON. I see you are back from class." Ellie liked Jon's good looks, but wasn't so sure about the conversation to come.

"Ellie" Jon started in, "I have been thinking about what you said when I passed this way earlier. It is really a question of whose authority you are going to accept as trustworthy."

"That's right," said Ellie, "and I don't think the American Academy of Science is very neutral or unbiased concerning the relationship between science and religion, and specifically Christianity. They are primarily agnostics and atheists, and have an axe to grind."

"Oh, come on Ellie, these are brilliant men and women of science. They know how to be objective even when they largely disagree with the bulk of the American populace. Remember 7 percent of them are theists, and some of those are Christians."

Ellie retorted, "But what kind of Christians are they; the wimpy kind that acquiesce in matters of nature, that science is always right and Christian theology had better get its act together and accommodate the latest scientific results into its theology or else?"

"Ellie, that is quite a slam on the Christians in the academy," replied Jon.

"Well, some may be the non-combative type, some too cowed and awed to speak up, some too old and tired to continue, and some bidding their time for the right circumstance to speak up" said Ellie. Jon looked at her condescendingly, "Nice personal opinion you have there. Joe of Podunkville thinks the whole academy is comprised of aliens disguised as humans who are trying to take over the country. Got any evidence?"

"Yes" said Ellie, "Bob Marks of Baylor University has not been invited into the National Academy of Engineering." "Who is he?" "Check out his credentials, pretty impressive; he should have been in there a long time ago."

"In your opinion."

"Well what about Henry Schafer III at the University of Georgia? He's tops in computational chemistry."

"Your opinion."

"Elected Chair of the World Conference on Computational Chemistry."

"So."

"Nominated for the Nobel Prize four times."

"And?"

"Fourth most cited chemist in the literature a few years ago."

"Oh, and why do you suppose he has not been elected to the academy," replied Jon.

"Because he thinks intelligent design is a valid scientific idea that needs to be investigated further."

Jon stared at her. "Do you really think they would keep a man of such scientific credentials out of the academy because of that?" He stated questioningly.

Ellie smiled back. "Well, it would give a tremendous amount of credibility to an idea that 90 percent of them don't like at all."

Jon paused and then stated, "You don't have a very high opinion of their ability to be objective in the science and religion area, do you?"

"Well, how else do you interpret actions like this?" Ellie quipped.

"Lots of ways," said Jon, "There could be lots of reasons on why he wasn't elected. Maybe he's some kind of Bible thumper."

"But why should Bible thumping discredit a man from the National Academy of Science? If he is good at what he does and gets the credentialing that he has, then maybe he isn't a Bible thumper, or maybe Bible thumping is irrelevant to the question of his election. And just what is a Bible thumper anyway? It is just a term we use to push somebody out of the conversation that we don't want in anyway"

"Ellie, this conversation is getting annoying." said a partially irritated Jon.

"Or discomforting?" said Ellie.

"Not really" said Jon, "You are still good looking" and with that, he walked off.

Ellie called out to him, "They are a community you know, with their own values and traditions." Jon turned around and shouted back, "Tradition, community, and science? You are losing it, Ellie!"

"Or finding it," she murmured to herself.

Hopefully at this point I have persuaded you that science and the scientific method are not well understood in terms of objective rules that can be written down and passed on from generation to generation simply by reading a book, somebody's paper, or lectures notes. That does not mean it does not work quite well or is not passed on from generation to generation. But it does raise questions about how it really works. How is it passed on? And how does science relate to the rest of life? For our discussion we really want to know; how does science relate to religious faith, and in particular, Christianity?

In 1946 the Hungarian physical chemist Michael Polanyi published a controversial and fascinating study of science entitled *Science Faith and Society*[1]. Polanyi was raised a secular Jew in pre-World War I Budapest, Hungary. He was a very bright child who studied at the best institutions of his day and country. In his early adult years, he converted to Catholicism, and later in life to Protestantism though he never formally joined a particular church or demonstrated a particularly ardent belief. He started his scientific career as a medical doctor, and then switched to physical chemistry after receiving a positive affirmation from Albert Einstein about one of his papers. Moving to Germany, he continued his career switch and ended up leading a spectroscopy group at the famous Kaiser Wilhelm Institute in Berlin. It was there that he established his reputation as one of the great scientists of his day and started on a career path that seemed potentially destined for a Nobel Prize. Adolf Hitler and World War II caused him to flee with his family to the University of Manchester in England to head up the physical chemistry division there. Shortly after the war his interests switched again to economics, and then to philosophy, where he made the amazing request to have his university appointment changed from natural science to the social sciences. Even more amazingly, the university president granted it to keep him at the university. This killed his chances of a Nobel Prize, but set him on a path perhaps even more enduring, (his son John Polanyi, did receive one in spectroscopy in 1986). Over the years since his switch from chemistry to social sciences, he authored several books that have become unusual classics in philosophy, epistemology, and the human condition. Polanyi was deeply troubled over why Europe destroyed itself with the disasters of

---

1. Polany, Science, Faith and Society.

WWI, WWII, and the fall of Eastern Europe to communism. The themes of freedom versus control, science, and knowledge as free enterprises versus its servitude to government, and knowledge as warranted belief versus empirical fact dominated his writings. He died in 1976, but quotations from his books and articles continually show up in philosophical and science-faith debates and writings. Some have argued that Thomas Kuhn got many of his ideas about the philosophy of science from Polanyi's writings and did not properly credit him[2]. Polanyi is one of four Hungarian scientists and mathematician who fled Germany during WWII who are jokingly called the extraterrestrials, because anybody that smart cannot possibly be human.

In *Science Faith and Society,* Polanyi attempts to understand science and how it operates within society. He starts by attacking the positivist notion of science as a precise procedure for finding truth. For example, take a collection of data that you can get for any phenomena: the position of the earth around the sun for any given day of the year, the density of water versus the temperature of that water, etc. Every good mathematician knows that an infinite number of functions will fit a finite number of points equally well. Most of these functions are going to be hideously complex. But often even a few simpler ones will do equally well in fitting the data at hand. Which one of these is the correct one? There is no scientific rule for choosing which function is the best one to use. Often the simpler function is chosen. But why is the simpler equation chosen over the complex one? If there are three simpler equations to choose from; which one should you choose and why? Natural laws frequently come from the functions found to fit the data. Then how do we derive the natural laws if we have no precise procedure by which to choose the equation? Polanyi's point: there is no precise procedure for finding natural laws. We guess the form of these natural laws. Science ultimately is a type of refined guessing. Not just any old guessing will do, we must guess, and ultimately our guesses must be shown to be true or as close to the truth as we can make it. Otherwise people die and very bad things can happen. Imagine trying to put a man on the planet Mars with a rocket ship using the earth-centered or geocentric view of the solar system and Aristotle's view of gravity.

But there is a fundamental difference between the guessing a scientist does and the guessing we often use in day-to-day life. In the middle of the night, you hear a dripping sound in the house. Is it due to the kitchen faucet not being closed properly or the pipes on the second floor of the house have sprung a leak? Is the water seeping through the second floor to drip in a puddle on your living room floor? A quick examination solves

2. Moleski, "Polanyi vs Kuhn".

the problem: you see water dripping from the kitchen sink faucet, or you slip on a puddle in the middle of your living room floor. But in both of these guesses you, are guessing at known phenomena that you have seen before and you know what sort of evidences these phenomena will give. A scientist is trying to guess at something that has never been seen or realized before. It is fundamentally new. An atom with a tiny nucleus, about which the electrons circle; a molecule that carries an informational code, DNA; all of these are new entities that may (or may not) have analogies with other completely different items (the solar system, a magnetic recording tape), but they are new and never "seen" until this point in time. How do you guess at something for which you have no previous knowledge? Of the myriad directions a scientist could take in her guessing and hypothesis making, why should she choose this particular route and why is it often successful? As Polanyi would say, "Whence can we guess the presence of a real relationship between observed data, if its existence has never before been known?" It is as if we have clues to a reality that we have never seen that *guides* us to its solution. What are these clues and how are we guided by them? A scientific discovery is like an arch of stones, all must be in place for any one stone to be stable, but in discovery you must put the stones in one by one. How does this happen? We must have some hidden foreknowledge of what and where the final arch is to make the discovery: we know more than we can tell. It is like a work of art where the artist has only an incomplete vision of the final piece, but in science we don't even know what the final work will look like. And the work is not one of our own making, but a truer picture of reality not within our powers of creating or shaping. *Thus we must know more than we can tell.* In *Science Faith and Society*, Polanyi likened this foreshadowing to a type of extra-sensory perception (ESP), but in later books and talks, he identified this foreknowledge with tacit knowledge. Tacit knowledge is a fundamental component of personal knowledge that he felt was at the basis of all knowledge, be it empirical or whatever.

If this foreknowledge is not ESP but a tacit component of knowledge then what is tacit knowledge and why is it important? Try to write down how you recognize a face. Do it with objective written out steps that will allow another person who has never seen your Mother to recognize your Mother using only the written description. They will have to identify her from a large crowd of woman the same age, hair color, and same clothing as your Mother, only the facial features vary. You will find it impossible to write an objective description of the recognition process that will work for anyone else. That is because the process has huge tacit components that cannot be verbalized or described accurately by language. Riding a bike or hitting a baseball with a bat has tacit components that can only be transmitted by example, mimicry, modeling, feel, and personal experimentation with the process over a period of time. Mentally assembling a sentence in your head and then speaking it has tacit components that defy written description. Polanyi argued that the process of choosing an experimental setup, research direction, interpreting the meaning of experimental results, and degree of persistence in achieving one's research ambitions before moving on to a new area, involves much tacit knowledge. There are no manuals prescribing the conduct of research because it could not be written if one even tried. The rules of research are like the rules of other higher arts, they are found in its practice. That is why professional researchers spend much of their graduate and post-graduate education in mentoring situations with master scientific mentors and teachers. It is the only way to pass on this tacit component of scientific knowledge. That is why master mentors often produce master students who go on to mentor others in the same grand scientific tradition.

A key component of Polanyi's writing can be found in his approach to scientific discovery. Scientific discovery is not the result of an impersonal, objective procedure that guarantees results, if one follows it. It is a mixture of empirical, personal, non-subjective, and subjective components that resembles a delicate personal art. Scientific discovery is usually a feat of emergence that does not come at the end of a long and intense period of mental concentration. But often it sneaks up on us when we are barely thinking about the matter, sometimes long after the agonizing mental battle to understand what is going on. Polanyi noted that many writers have noticed a pattern to the creative rhythm of discovery. It is often split up into four phases: Preparation: the coming up to speed on the details and background of the problem, Incubation: the time spent in thoroughly immersing oneself in the experimental or theoretical details of the issue and pursuing answers and guesses of your own, as applied to the problem, Illumination: the answer comes to us. Polanyi describes this as follows:

> Our labours are spent as it were in an unsuccessful scramble among the rocks and in the gullies on the flanks of the hill and then when we would give up for the moment and settle down to tea we suddenly find ourselves transported to the top.[3]

And finally there is Verification, where the investigator goes back into the lab to justify and support his guessed answer with empirical data, or an analysis of previous data that demonstrates the illumination. Scientists and students of many fields have had this experience. My own discovery experience occurred during my PhD dissertation work in physical chemistry. I was

3. Polanyi, Science, Faith and Society, 34.

trying to deposit a thin film of pure heavy water (water, D2O, with two deu-teriums instead of hydrogens) on a very cold surface in a vacuum chamber. If any normal water (H2O) vapor was present, the heavy water deuterium would exchange with the hydrogens to form a partially deuterated molecule (HOD) that would mess up the experiment. I struggled for a year and a half trying to exclude the normal water from the icy deposit. Fearing my disserta-tion research would fail, I was desperate to try anything that might work. After months of beating against the problem with unsuccessful deposit after deposit, I found myself with a notion to wrap a copper shield around the cold surface that was cooled to the same low temperature. It had a smaller opening that allowed the heavy water vapor to come in from the direction of the heavy water vapor nozzle. To my great surprise, the shield worked marvelously, and the research proceeded successfully. What is intriguing to me personally was that the idea came after a period of almost complete sur-render to the problem and after considerable prayer on my end, since I am a Christian who prays. To this day I am not completely sure why the shield worked, though my guesses are recorded in the dissertation. Many agnostic researchers have had similar experiences, but without the prayer.

Polanyi pointed out another very similar discovery experience that is fascinating and rich with implications.

> The solution of riddles, the invention of practical devices, the recognition of indistinct shapes, the diagnosis of an illness, the identification of a rare species, and many other forms of guess-ing right seem to conform to the same pattern. Among these I would include also the prayerful search for God. The report of St. Augustine of his long labours to achieve faith in Christianity, abruptly culminating in his conversion, which he immediately recognized as final and followed up by the lifelong vindication of the suddenly acquired faith, certainly reveals all the charac-teristic stages of the creative rhythm.[4]

This is a stunning comparison to make to the modern mind that of-ten refuses to see a connection between scientific creativity and religious faith. However, to the typical medieval academician the connection once explained would have been obvious. Why is our culture so different? And in this respect, which culture is right, the medieval or the modern? At the risk of being contrary I suggest the medieval is closer to the truth in this instance. Polanyi then states that "all of these processes of creative guesswork have in common that they are guided by the urge to make contact with a reality, which is felt to be there already to start with, waiting to be apprehended."

4. Polanyi, Science, Faith and Society, 34.

Michael Polanyi and Thomas Kuhn had much in common in their views of science, but here the two scientific philosophers depart in radically different directions. Kuhn was not convinced that the scientist was guided by the urge to make contact with a reality; Polanyi was. Martin Moleski[5], Polanyi's biographer, lifted these comments from Maben Walter Poirier to illustrate these radically different directions.

> For Polanyi, truth in general, and in the natural sciences in particular, is understood to be a fundamentally correct insight into the real, as it is independent of human thought processes. . . . Truth, for Polanyi, is not to be found in the collective aspirations of the community of scientists, or of its leading members, as seems to be the case for Kuhn. It resides in the judgment of a scientist, who, because of his fell for a particular subject, correctly claims that here is the real. . . . Polanyi is a philosophical realist, and not a radical relativist like Kuhn.[6]

This book is taking the direction of Michael Polanyi, not Thomas Kuhn. As a Christian I am committed to the concept that underneath reality is the real and it is true; and that it is very possible to approach and understand it. This concept is ultimately a faith commitment; or if we use the jargon of philosophy, a *presupposition* we have about the world. The religious overtones are becoming apparent at this stage. However if one were to take a more Kuhn-like position, and argue that science is like a type of biological evolution that wanders from one species to another with no goal or purpose in mind, i.e. one darn thing to another, then one has made the presupposition maybe there isn't any true or real underneath reality. This too is ultimately a faith commitment. One way to evaluate competing worldviews is to examine their internal consistencies and follow through what results from being internally consistent in your worldview. Polanyi far surpassed Kuhn in developing that internal consistency, and projecting what it will mean for the future of science. In my opinion, it is because Polanyi's initial *presuppositions* are much closer to the truth and hence his analysis and predictions ring much truer and match up with our scientific experience. If the driving force of a scientist is "the urge to make contact with a reality, which is felt to be there already," then how is this demonstrated in the practice of science?

We can support the idea that intuition and discovery are aspects of truth waiting to make it itself real in our minds with the curious cases of coincident discoveries. Polanyi listed several. Both forms of quantum mechanics Heisenberg and Born versus Schrodinger arose at almost the same

---

5. Moleski, "Polanyi vs Kuhn", 18–19.

6. Poirier, "A Comment on Polanyi vs Kuhn".

time. In 1923 de Broglie suggested electrons have a wave nature. Later in 1925 Davission and Germer, not knowing of de Broglie's theory, found that electrons "diffract" through a crystal. Paul Dirac's relativistic prediction of a positron in 1928 was confirmed experimentally by Anderson in 1932; who had no knowledge of Dirac's work. Yukawa predicted the existence of the sub-atomic meson particle in 1935, and it was discovered by Anderson in cosmic rays in 1938. I have my own story of coincident discovery[7]. A well known computational chemist Dr. Peter Pulay had thought of a way to calculate Raman polarizability derivatives by finding the derivatives of the computed molecular energy in the presence of an imposed computational electric field. As he started to work on the method, he was asked to review a journal paper which listed and explained the exact method he had envisioned. Being an honorable man, he dropped his paper idea and gave full credit to the group that beat him to the idea which is now known as the Kormiki-McIver algorithm. Could it be that the same intuitive contact guided these different scientists to the same hidden reality in all of these coincident discoveries? Perhaps it is more than *I know more than I can tell*; it is the *community of science* knowing more than *it* can tell. The story of Pulay being scooped on the discovery of the Kormiki-McIver algorithm and his response to it illustrates another very curious component of the scientific enterprise—the moral component.

"Bruce, I think I have an idea that will greatly improve the stability of the argon ion laser I use to collect my Raman spectra with." said Bill.

"What is it?" asked Bruce.

"Well, I want to put a microscopic slide into the laser beam and reflect off a very tiny fraction of the laser light to a photodiode whose electrical output will be coupled back into the laser power supply light amplitude control input. When the beam intensity drops, the electrical current will drop and boost the light amplitude back up and do the opposite if the beam intensity increases, thus keeping the laser light amplitude at the same light level continuously."

"Well if you think it will work, give it a try, but don't spend too much time on it," replied Bruce. Two days later Bill had the whole setup going and was busy examining the laser output data with a dour eye.

"Is it working?" asked Bruce as he wandered into the lab where the laser was busy humming along.

7. This story was told to me by a Hungarian scientist friend, Dr. Gabor Pongor who worked extensively with Professor Peter Pulay.

"Well, it doesn't seem to be decreasing the noise any, it actually seems to have made matters slightly worst. Maybe if I changed the circuit a bit I could get improvement." Bill replied.

"Bill, you have to be hard on yourself when it comes to evaluating data. If the results are not better than what you started with, then maybe you have to concede that the idea is wrong, and that you need to move on to more important things." Bruce intoned as he wandered out the lab door.

This was not what I wanted to hear. I was a young scientist anxious to prove my worth at my first professional research assignment, and Bruce was my boss and group leader. The sting of that failure is still tangible today. You would think that an objective truth-seeking scientist would be delighted to be shown that his work was heading down the wrong path: joyful that his error was exposed before he published it to the scientific world, leading them on the same path of error. But no, the primary emotion is bitter disappointment and confusion. Why? To anyone who has ever invested extensive time into a task only to see it fail, the answer is obvious. But suppose the scientist is on the cusp of a smashing success that will change his field and boost his reputation up to the best scientists? One more try and he is there. Or suppose he is chasing a useless research path, one that will never yield good results; results that he could find quickly if he changed to another approach. Then what? Polanyi remarked that a scientist is constantly weighing his research convictions and faithfulness to an ideal of scientific conduct with honest self-criticism and care;

> the scientist may appear as a mere truth-finding machine steered by intuitive sensitivity. But this view takes no account of the curious fact that he himself the ultimate judge of what he accepts as true . . . the scientist is detective, policeman, judge, and jury all rolled into one.[8]

What guides the scientist is his scientific conscience. It stands apart from the individual scientist like a judge that arbitrates the continual decisions and mini-crises he faces in his research. You could speculate like mad, and see every wishful conclusion you wish in your experiments if you tried hard enough. Or you could bring the full weight of your critical analysis to bear on every experiment, abandoning them and every hypothesis after the slightest failure. The former path is reckless and could potentially lead you to the Nobel Prize. It could also make you look like a fool to your scientific peers. The latter path of ultimate critical caution could so paralyze your

---

8. Polanyi, Science, Faith and Society, 38.

work that you spend years going nowhere. What guides a scientist down the middle path? His scientific conscience. That can only be explained by looking at the moral elements of science. Explaining the moral elements of science empirically or with any proposed scientific method is a lost cause. Summarizing, how do you explain the scientific method scientifically? You cannot, and it is a ridiculous question to even ask.

"Let me get this right, you say science has a strong moral element that cannot be justified by any appeal to science." Mary looked genuinely puzzled as she gazed at Ellie. She and Ellie were casual friends, and Mary was intrigued and sometimes amused by Ellie's strong statements and constant referencing to God. Not that she was any agnostic or anything, but how can one really know anything about matters this complex?

"Yes" said Ellie.

"Dr. Swartzmann would certainly not agree with you."

"I am not in that class; what do you mean?"

"He believes that morality can ultimately be explained as a long term result of the Darwinian struggle for survival. The species' genes are the basic units that fight for survival, the animal species itself is just a machine that the genes wrap around themselves to facilitate their survival. This all happened over millennia, he claims. Thus cooperation and altruistic traits such as selflessness and self-sacrifice for others are just the result of genes that give in lesser ways to gain overall in the long-term strategy of self-survival of the basic biological unit, the gene."

"Can you simplify that a bit?" gasped Ellie.

"I'll try. Species genes are always competing against each other in the struggle to survive. Sometimes two or three genes (species) will stumble upon a life scheme that really looks like giving to each other when it is really a scratch my back and I'll scratch yours in a common plan to beat the environmental odds that are out to get them both. So if they happen upon an altruistic plan between them that overall improves their chances for survival, then they will out-compete the other competition in the game and life and survive better and take over their environmental niche. He claims human altruistic behavior can be explained by that model. So things like values and morals can be explained by evolution. Since evolution is a random purposeless process, then values and morals are deterministic processes resulting from evolution."

"And then?" asked Ellie.

"Well, the moral component of science is explained by evolution which is science. So science has explained the moral component of science, and doesn't need a religious answer," replied Mary.

"Do you believe that?"

"Not really" answered Mary, "But what is a good reply to such an argument?"

"Well, I can think of a few."

"What are they?"

Ellie paused and thought a bit. "Well, first we really don't even know what a gene is really; fundamentally it is just a biological concept."

"But I thought a gene was a strip of DNA that codes for a particular protein or series of proteins that gives an animal or organism a specific trait, right?" explained Mary.

Ellie replied, "Remember the gene knock-out experiments, you knock out what you think is a gene for one trait and end up affecting three other traits and only partly affect the trait you are after. In short, just what is a gene? It still is a very abstract idea in higher organisms."

"Okay" said Mary, "How does that affect Dawkins' ideas of the selfish gene?"

Ellie was quick to reply, "Well it does beg the question as to how a strip of DNA or some complicated chunk of DNA and other biochemistry in you is going to cause you or me to give up reproducing, by becoming a nun, or sacrifice our lives to save someone else's child. How does something as me-chanical sounding as a gene affect the entire population of humans so that we are nice to each other, because down deep we all want to propagate the human species genes? That is one heck of a gene. Maybe we all ought to bow down and worship the god-like entity that has such preeminent power over each one of us individually every day. How does something like selflessness emerge out of our biochemistry? Molecules to social consciousness; where are the details other than saying we have not figured it out (and may not); but it still must exist. Because I say so?"

"Hey, there is Bert. He is a philosophy major, let's ask him," interjected Mary quickly. "Bert! Come over here! Tell us what is wrong with Richard Dawkin's selfish gene idea as an explanation of altruism and morality."

Bert did not match the stereotype of the long-haired hippy philoso-pher. He looked more like a CEO or business major who dressed just a bit causally; who thought a bit before he spoke. Otherwise you would have picked him for a well-dressed business student, and he strode over much like his demeanor predicted. "Dawkins, selfish gene; what brought this topic up?" Bert stated.

"Dr. Swartzmann's class," replied Mary.

"I should have known," sighed Bert, "The guy has no life or brains outside of his academic life."

"Yes he does!" exclaimed Mary, "His live-in girlfriend Charlotte!"

"Figures," said Bert.

"Well what would you say to Swartzmann or Dawkins?" asked Mary.

"First you ask, why did the concept of selflessness and sacrifice ever come to be? And why did it come to be in a world determined by genes with the sole purpose of self-propagation and selfishness. Selfishness is a term of abuse in our world, entirely negative. If self-interest is at the core of all human motivation, then how did such a word, selfish, get invented? Why such words as mercy, loyalty, and justice? Why are they considered good? By the way, didn't Dawkins argue that we should rebel against our genes and develop a kind, good, altruistic world?

"Yes! That's right." said Mary, "He argued that we have evolved to the point that we can choose to not obey our genes and create a better world, and that we humans alone have the ability to do it."

"Why should we?" replied Bert, "By what standard does Dawkins appeal to say that it is better for us to rebel against our genes? But if selfishness is in my best interest, then why would I want to rebel against my genes?"

"Wait a minute!" cried Ellie, "Bert, are you saying that Dawkins is implicitly appealing to some kind of higher universal standard by even implying that we ought to rebel against our genes?"

"You got it."

"And by trying to explain away the problem of selflessness in a world supposedly built totally on self-interest, Dawkins and Swartzmann are unconsciously admitting that this unexplainable standard of good and bad exists: when really it should not even arise in a world built by self-interest." explained Ellie, "This is C. S. Lewis's law of human nature all over again."

"Yup" replied Bert.

"Ooo, this would be fun to try on Dr. Swartzmann." exclaimed Mary.

"Don't be surprised if he makes appeals to the ability of natural selection and time and the complexity of human social organization to pass off the universal standard of morality argument as an evolved mental construct. If we knew evolution well enough, it could be explained by evolution. If you start with a totally materialistic worldview, then natural selection, evolution, or some material process *has* to be able to do the impossible. There is no other explanation, by fiat." countered Bert.

"Keeps the guilt away, right!" sneered Ellie.

"Maybe so." replied Mary.

Bert just stared and then strode off. "See you later" he intoned.

Science is a community, a society, a type of democratic institution with its own set of rules, traditions, leaders, apprenticeships, morals, beliefs, faith, and spirituality. If I had said this at the start of this book, you may have likely thrown it down in disbelief; particularly if you are a practicing scientist. But hopefully we have wrecked the simplistic idea of science as a fail-safe procedure for finding truth in a straightforward objective fashion. Let us pick Polanyi's arguments apart further and see if this "science is a community" sentence can be justified.

Learning to think scientifically, or becoming a scientist is not an overnight decision; it is a process. First science is taught in elementary and high school in a doctrinal way. The classic Baconian method of induction with data collection, induction, hypothesis, experiment, new hypothesis is taught as the letter of science. It's dead and not representative of what really takes place in science, but at least it is a starting point. Theory and scientific knowledge are presented as established facts with the full approval of the scientific establishment. This is as far as most people get in their initiation into science, and hence the massive public misunderstandings of the scientific institution are understandable. At the university level, the aspiring science student learns the methods of science better. They may encounter a scientist who has practiced scientific research as a profession and published among scientific peers. He learns the provisional nature of science and some of its uncertainties. For the first time, he sees the possibilities of science and starts to envision his own role in the scientific community. If he persists in his chosen scientific field, a privilege that is granted only a few outstanding students, then he is fully initiated into the profession by a mentoring apprenticeship relationship with an established master of science. The graduate school environment oversees this mentoring process, and ensures that it was done with proper academic rigor. This "apprenticeship" with a science master greatly influences the student and colors the rest of his scientific career. He learns his mentor's speculations, approach, opinion of other scientists' work, and his mentor's grand vision of what science is and how it functions. Much of what we call the practicing of scientific methods is not so much taught as it is caught by the student. He learns by doing, watching, seeing the successes and mistakes of others and formulating his own slightly unique approach to his subject. He is constantly asked to demonstrate his knowledge of the basic material and to justify why he researches as he does daily if not sometimes hourly. Thus much of the practicing skill of scientific research is taught as an art, or as tacit knowledge. That is why it is so hard to objectively define the actual intimate workings of science. Much of the knowledge learned is objective and easy to record, but much is also tacit, known, and obvious, but not explainable objectively.

How well does this mentoring work? Sometimes stunningly well. Professor Ernest Rutherford, the discoverer of the atomic nucleus, won the Nobel Prize in science for his scientific work with atomic physics. He mentored many graduates and post-graduate students. Four Nobel Laureates can be found among his personal pupils. Michael Polanyi likened this chain of brilliant scientists to a modern version of apostolic succession. The apostles and the early Christian church fathers trained and mentored their disciples in God's special revelation—Jesus Christ, the Holy Scriptures, and their own personal experiences with God and the Holy Spirit. Rutherford and his students/disciples were trained in God's general revelation, the physical world, with the general revelation it contained, and their experiences in trying to comprehend it. This is surprising; especially when many of Rutherford's students may have rejected God's special revelation, and did not perceive the natural world as containing God's general revelation as all. But the discipleship, tradition, vision, and faith that drove the Rutherford succession is there anyway, and this succession line of scientists would have faltered without it.

What drives a scientist's vision and faith? It is his belief that the parts of scientific knowledge that are not understood or discovered yet, are true and valuable; worthy of investing an enormous amount of one's time. Polanyi offers a quotation worth repeating.

> A child could never learn to speak if it assumed the words which are used in its hearing are meaningless; or even if it assumed that five out of ten words so used are meaningless. And similarly no one can become a scientist unless he presumes that the scientific doctrine and method are fundamentally sound and that their ultimate premises can be unquestioningly accepted. We have here an instance of the process described epigrammatically by the Christian church fathers in the words: *fides quaerens intellectum*, faith in search of understanding.[9]

To Isaac Newton and his peers this would not have sounded unusual. Today it sounds scandalous. True, the faith of the average top tier scientist is certainly not guaranteed to be a Christian faith, but it is a faith in their ideal of science, and it is pursued with equal vigor.

As the science student grows and matures he will slowly rely on authority less and less for their scientific beliefs and vision and more on his own judgment; particularly in areas where they are becoming experts and have been exposed to an array of scientific opinions. The graduate student's submission to his mentor's ideas will slowly decline as his own experience and thoughts gain in confidence and particularly when his mentors acknowledge his correct interpretations and vision in science in the areas where he is becoming an expert. The good mentor acknowledges and

9. Polanyi, Science, Faith and Society, 45.

encourages this scientific independence in his pupil, even to the point of being shown wrong by his student, precisely because the mentor believes in the community of science and its ability to ascertain truth. And he wishes for his student to join that morally grounded community as a full-fledged member with the same degree of scientific independence that was granted to the mentor when he started.

"This is not going to be an easy meeting." thought Charles as he wandered into his graduate advisor's office that morning. Dr. Melvin his graduate advisor, with another faculty, had just published a quick note to the Journal of Chemical Physics Letters on excited states using a laser method where two laser beams crossed to produce a third generated beam that would contain spectral information. The results looked exciting and promised a whole new area of spectral research and funding. However Charles had been charged with verifying the results and seeing if he could push the method a bit further as a possible thesis project. He had, and what he had found out was not what he had expected. A recent monograph on the method showed that the crossing angle of the two laser beams was critical and that the results varied wildly with the crossing angle of the two input laser beams.

"Well Charles, what have you got for me?" asked Dr. Melvin.

"Dr. Melvin, I could never quite get that high rising peak that you got at the detector" replied Charles.

"Did you get partially close to it?"

"Yes many times, and I could get it to transform to a valley or flatten out to nothing also."

"You could!? How did you manage to do that? Was it reproducible?" questioned a very inquisitive Dr. Melvin.

"Well I had read in Truccia's monograph how there was a close relationship between the two laser beams crossing angles and the shapes of the peaks you see at the detector output. See here it is in Trucca's paper, here are his calculated peak shapes, and here are the peak shapes I got when I varied the crossing angle between the two laser beams." Charles quickly placed Trucca's paper and his own data in a pile before Dr. Melvin. Dr. Melvin rapidly scanned the paper and then stared at Charles's data graph by graph, with intense concentration.

"How did you vary the crossing angle?

"I put a small lens in front of the laser beams and rigged it up with a screw I could turn that moves the lens up and down the two input beam axis. Move it up the angle increases, down it decreases" replied a nervous Charles. "Move it up you sort of get a peak shape like you and Dr. Grassely got, move it down and it disappears"

"So do you think that the peak Dr. Grassely and I got was ju
fact of tuning the lasers up and accidently changing the crossing
not a real excited state?" questioned a slightly irritated Dr. Melvin.

"Dear God!" Thought Charles to himself, "How am I going to get out of
this one? Here I am his graduate student telling him his note to the journal
is probably junk! Who am I to tell him this, how do I know I got it right?"

"Well, you never really know about these things, it may be possible
to tune the lasers up and not change the crossing angle. I am certainly not
as good or patient as you in getting those lasers to work right" breathed a
worried Charles to Dr. Melvin.

"Hmm" said Dr. Melvin, as he shifted back in his chair eyeing Charles
with an unusual look, "Let me think about this a while." Six months later
as Charles was working on his new and different thesis project Dr. Melvin
called out to Charles, "Charles come on down to my office for a moment will
you?" As Charles wandered in and sat down, Dr. Melvin started, "Charles
how is the FTIR work coming along?"

"Fine," replied Charles.

"Well I ran across this paper in the *Journal Chemical Physics* that referred to Dr. Grassely's and my note on that laser method for excited states."

"What did they think?" mumbled Charles.

"They didn't think very much of it" stated Dr. Melvin.

"Oh."

"Can you think of any reason why they should be wrong in what they
said?" asked Dr. Melvin.

"Uh, not really sir" replied Charles very faintly.

"There was an awkward pause and a slight sigh as Dr. Melvin said,
"Okay, thanks Charles, hurry up on that FTIR data."

"Amazing," thought Charles as he walked back to his work. "Had that
been Dr. Paris, he would have skewered me for contradicting his research
conclusions. Boy, did I choose the right mentor this time."

Dr. Melvin was leaning back in his seat pondering his conversation
with Charles, "He actually suggested I was wrong. Maybe we'll make a scientist out of him yet."

Without this belief in a higher authority, that truth is more important
than one's personal immediate success, it would be difficult to pass on the
traditions and values of science. Granted no one scientist is perfect in this
ideal, but it is an ideal that is held above each scientist, not only by himself,
but by the community of scientists about him.

What are some of these values that the budding scientist is assimi-
lating from his educational scientific community? There are moral values.

You should be honest in reporting your data; you should be accurate and not sloppy in your experimentation and data collection; and you should be willing to share your data with everyone who asks within the community. For science to advance, everyone needs access to the data that others have produced. Thus scientists are encouraged to publish their data in the public domain for others to evaluate and use. It is understood that scientists have a right to withhold data until they have verified it internally, published it, and established their claim as the discoverers and first interpreters of its meaning. But after this point, refusal to publicize the methods and key data raises tremendous suspicion, mistrust, and doubt among the community of science.

A classic example of this was demonstrated by the cold fusion furor of the early 1990s. Two scientists working at the University of Utah claimed to have produced the fusion of hydrogen atoms, (actually heavy hydrogen or deuterium), into a helium atom, releasing an enormous amount of heat energy as an expected by-product of the atomic fusion reaction. The discovery was announced with great fanfare by the university prior to publication. A new day of energy production was at hand. The method was surprisingly simple; place heavy water (D_2O) in an electrochemical cell with palladium electrodes and apply the correct voltage and current to the electrodes. Scientific groups and energy companies everywhere were scrambling to replicate the details of the experiment from the few details that were leaked out. Months down the road, the expected paper was not published and attempts to reproduce the fantastic results failed. The clamor from the scientific community for details and more explanations rose furiously. I remember conversations where in exasperation my fellow scientists exclaimed, "Why don't they just come out in the open and publish all their results? They have established their claim to the invention. What are they trying to hide?" Eventually the results were published in the journal literature but the claims considerable muted and the data severely lacking in quantity and quality. As expected the scientific community quickly abandoned cold fusion research and the reputations of the original scientists were tarnished, if not ruined.

Scientists also use aesthetic values. Ockam's razor; the idea that the simpler of several equivalent ideas that fit the data should be used, has its origin in aesthetic values, the value of simplicity is better. Beauty in an equation or theoretical model is greatly admired in science. Werner Heisenberg, Nobel Laureate, and founder of modern quantum mechanics, was quoted in a conversation with Einstein.

> You may object that by speaking of simplicity and beauty I am introducing aesthetic criteria of truth, and I frankly admit that I

am strongly attracted by the simplicity and beauty of the mathematical schemes which nature presents us. You must have felt this too: the almost frightening simplicity and wholeness of the relationship, which nature suddenly spreads out before us . . .[10]

Scientists presuppose knowledge-based values. A theory should have predictive success. It should be internally clear and coherent in all of its explanations. Scientists accept without much question methodological values. Double blind experiments are better than single blind experiments. Two point calibrations of instruments are greatly preferred to single point calibrations. Data points that differ from the average by over three standard deviations are highly suspect. How do you empirically or rationally justify such values? Can you scientifically prove them to be the best values to base science on? No, and no one even tries, you just assimilate them as a budding scientist and move on. Here a telling quotation from Polanyi that illustrates the implicit value system of honesty and integrity; which without the entire structure of science would completely collapse.

> The quickest impression on the scientific world may be made not by publishing the whole truth and nothing but the truth but rather by serving up an interesting and plausible story composed of parts of the truth with a little straight invention admixed to it. . . . If each scientist set to work every morning with the intention of doing the best bit of safe charlatanry which would just help him into a good post, there would soon exist no effective standards by which such deception could be detected. A community of scientists in which each would act only with an eye to please scientific opinion would find no scientific opinion to please. Only if scientists remain loyal to scientific ideals rather than try to achieve success with their fellow scientists can they form a community which will uphold these ideals.[11]

What about the traditions of science? The values just described are a part of the grand tradition of science, but it goes further. For example how is scientific information produced, accredited, and disseminated? There are lots of reasonable ways it could be done, but science has its established tradition, a disseminating tradition that has worked well. First the data is collected, analyzed, and then submitted for publication in a reputable scientific journal. At the journal level the paper is sent out for peer review, where one to three experts in the field are chosen to review the paper, and recommend publication, publication with modification, major reworking

10. Davies, God and the New Physics, 220–221. Davies is citing Heisenberg.
11. Polanyi, Science, Faith and Society, 54.

of the paper, or outright rejection. The editor of the journal selects the re-viewers and manages the whole review process. In earlier years like when Einstein first started in 1900, the journal editor himself would review the paper and decide what action would be taken. At government agencies the individual scientist will write reports that are compiled by the group manager in a report that will be published by that agency. The methods vary, though peer review is the most common for most journal articles. But in all cases either a senior scientist or another fellow experienced scientist who is familiar with the field will review the work and decide whether it is suitable, and up to publication standards. Once it is published, the scientific community scrutinizes the paper. It can be hailed as a new break-through, important, just interesting, or plain pedestrian. If it is accepted and consid-ered important then it moves on to become part of the review literature, and eventually makes its way into scientific textbooks. In critical foundational areas, this process can take a very long time. The field of modern quantum mechanics was developed from 1926 to 1928. It was not until the fifties and sixties that freshman general chemistry textbooks started explaining atomic structure using detailed quantum mechanical explanations simplified to the university freshman level. Polanyi described what the community of science is looking for in these important papers. Is it valid over the areas of sci-ence about which they are concerned? Is it profound; does it open up whole new possibilities to discover and develop? Is it of intrinsic human interest; why should they care about it? If it is, then the community of science will embrace it and use it. This helps explain the curious result that even an iconoclast scientist who is trying to sell a completely different theory than what is accepted will greatly value the opinion of his fellow scientists, and try his best to persuade them to his point of view. If he succeeds, he gains credibility; if not, he loses credibility. How is this, a tradition? Let's examine some alternate methods or traditions of accrediting scientific work.

First select another way of choosing which papers get published. Suppose we roll dice or randomly choose which papers get published. The resulting journal would be terrible. Cranks, irrelevant work, trash, and all sorts of nonsense would be published along with the occasional really good work that really did show something new and trustworthy. The good stuff would be crowded out and lost in the junk. The inexperienced scientific reader would be unable to distinguish between the good and the bad, and quickly become discouraged and confused. This is exactly the situation we have with a typical search on the internet. The lies, conspiracy theories, confused, and inaccurate is mixed in with the top notch and competent. It takes a highly trained mind to have a chance of distinguishing it all. The modern internet is truly a wonder of useful information, but it is a wonder of misinformation also. A good quality scientific journal removes the trash component so that reading and searching effort is maximized, and targeted on the hopefully true. Fortunately scientific tradition has not chosen such a lousy random vetting method but sustained the previous method over centuries.

Suppose the presidents of academy of sciences of every country banded together and decided that they alone would be the final arbiters of every scientific question. If they and their underlings did not approve a paper, it would not get published. Their decisions would not be respected and the whole community of science would grind to a halt. No scientist with a love of science would join such an institution. Scientists expect to have some say in who governs them. They cannot respect or submit to those whose opinions they do not value. It is the grand tradition of science, and bears a fascinating resemblance to the representative republic form of governance. It is a tradition that a scientist swears allegiance to, believes in, and abides by. Polanyi phrased it well in this quotation.

> It would appear that when the premises of science are held in common by the scientific community each must subscribe to them as an act of devotion. These premises form not merely a guide to intuition, but also a guide to conscience: they are not merely indicative, but normative. The tradition of science, it would seem, must be upheld as an unconditional demand if it is to be upheld at all. It can be made use of by scientists if they place themselves at its service. It is a spiritual reality which stands over them and compels their allegiance.[12]

This is a stunning quotation if we grasp its full significance. Devotion, intuition, conscience, tradition, service, allegiance, and spiritual reality: who would initially ascribe these qualities to the competent practicing scientist? But they must be there or the very enterprise of science itself is threatened. But many human institutions have these traditions, qualities, ideals: republican forms of democratic government, rotary clubs, local governments, universities, churches, church denominations, legal bar associations, much of western civic and professional life is wrapped up in this same grand type of tradition. Polanyi expresses it this way, "scientific opinion, legal theory, Protestant theology are all formed by the consensus of independent individuals, rooted in a common tradition[13]." That is the whole point of this chapter. *Scientific institutions are not very different; they are basically the same.* So, why is scientific knowledge placed so high above other forms

12. Polanyi, Science, Faith and Society, 54.
13. Polanyi, Science, Faith and Society, 57.

of knowledge in our culture, particularly theological knowledge? On what legitimate grounds of experience do we do this? It deserves a high place, but how high; and has our western culture overdone it at the expense of other valid forms of knowledge and ways of knowing? I would not be asking the question if I did not feel that science has gone too far.

"Ellie! Are you out of your freaking mind!" exclaimed Jon.

"No, I am quite sane, thank you" said Ellie. "And I will stand by my belief that scientific institutions and science itself is not that fundamentally different from Protestant seminaries, denominations, and the legal bar associations"

"Ellie that is quite the claim to make" said Mary.

"Well you have heard my arguments, devotion, tradition, conscience, and allegiance to ideals that cannot be justified by any empirical means drives science just like it does Protestant theology and the legal profession.

"But Ellie, why are you excluding the Catholics?" questioned Mary.

"Well, I am not" said Ellie, "But there is a big difference between how Protestants view theological authority and how Catholics do. Catholics vest their ecclesiastical authority in the Pope and the College of Cardinals at the Vatican in conjunction with the Holy Scriptures. The decisions of how a Catholic should believe are made at the top in fairly extensive detail and interpreted by the local priests for the congregation. Protestant theology is less rooted in past tradition and places more emphasis on the Bible as it is read and understood by the leaders in the denomination. A more generalized set of beliefs is handed down from the seminaries and assemblies of the denominations, asserting the basic Christian beliefs. Then individual Christians are given the freedom to interpret specific scriptural passages for themselves under the overarching set of basic beliefs handed down from the denominational hierarchy. Some Christians prefer one authority structure, Catholic; others prefer another, Protestant. But science and its institutions more nearly resemble the Protestant traditions in its workings. That's all I am trying to say."

"But Ellie, scientists are not bound by the belief in God or a divine book of scriptures, they can go where they wish in their scientific research" replied an exasperated Jon. "We just use our reason and rationality to investigate the world.

"Jon, scientists are too bound; and very bound up in where they can go in their research. Try to investigate the possibility that an intelligent agency injected information into the living world in the very distant past and see how far you can go and get funding, or present in a professional science meeting. You can barely crack the door with such an idea."

"But Ellie, that is religion, not science! Save that for your religious meetings and theology conferences. Don't mess up science with half cracked ideas of angels dancing on the heads of pins, and creating dinosaurs!" exclaimed Jon.

"Ellie, Jon does have point there, what are you trying to say? You just can't open the door of science to just any supernatural occurrence or scientists could start claiming, God here, God there for anything they can't exclaim rationally." murmured Mary.

"Okay, I am not saying throw the doors of science wide open to anything, just that we need to reexamine the basic presuppositions of science, the unspoken ones no one ever really discusses, to see if they really fit the natural world and the data we see today" countered Ellie.

"And what is that?" stated Jon a trifled amused.

"That the forces and particles did it all."

"What do you mean by that?"

"Well" said Ellie, "that is a catchy way of saying methodological naturalism, the belief that the universe is a closed system, and does not need any outside agent like God to explain how things came to be or operate. All we have to explain the complexity of the universe is the force, laws, and basic particles that were present in the universe after the big bang creation of the cosmos. Try a thought experiment with me"

"What do you mean?" said a rather tart Jon.

"Grant me for the sake of argument that God really did intervene in some special way in the past history of the earth. How are you ever going to explain part of that history that God worked with if you are philosophically forced to say the forces and particles did it all? You will never get the right answer; just some just-so stories that don't add up. To explain an event that includes the action of God you must have as a starting presupposition the possibility that an intelligent agent was possibly involved. You are not forcing God to be involved, but just possibly"

Jon quickly answered back to Ellie. "But suppose you say there is no God and you can explain everything reasonably well, or you suppose God did something and it turns out there is a perfectly good natural explanation for it."

Ellie shot back, "If you find an acceptable reasonable material explanation it just means God's action is hidden or was not present as a primary cause in that event, but it doesn't rule him out of future events."

"But Ellie you are acting like God's actions can actually be put to the scientific test somehow. You can't scientifically prove God" rasped an astonished Jon.

"Jon, science doesn't prove anything; all it does is make statements of increasing and decreasing accuracy" retorted Ellie.

"You are absolutely right" said Bert. Everyone turned to stare at Bert who had just walked up to the café table where the threesome was sipping sodas on that bright sunny day. "It is really a matter of degree, you build your case for or against something until it somehow reaches the point of change, when your epistemological means of knowing become convinced that this is where the truth is, and you change your mind or become more firmly rooted in your belief." Bert eyed Jon with a mischievous grin and then fished around in his pocket for some spare change. "Give me a Coke please" he told the vendor as he snapped his money down on the kiosk counter.

"Jon, there really is no such thing as absolute scientific proof" exhorted Bert. "Some scientific statements are very reliable, others less so, the good scientist is driven by a faith in his perception of truth, much like a good Christian. The question is whose overall perception of reality is closer to the truth: who corresponds to reality better. The two don't have to be exclusive of each other also. The good scientist could be the good Christian. The real test of will is between the philosophical materialist and the reality influenced by God, theist."

"Two worldviews, aye, Bert; the materialist view of the universe versus the theist view," yawned Jon. "We have been fighting the dark side of spirits, ghosts, and goblins since the enlightenment." With a side glance at Mary, Jon stated, "You don't have to give up believing in God to embrace the rational view of the world. You are right, no one can really disprove whether God exists or not. God could have set the whole thing up and set it running somehow so that we could never detect his action, or maybe God gave us reason and logic as the sole way to figure out the world and make sense of things. You have got to admit, that my way of looking at the universe is the dominant way the university explains the world. Maybe God exists, but who really cares unless it helps you cope with the realities of life."

"It wasn't always this way in the West, and a lot of people are getting pretty fed up with the shallowness of the scientific materialist view of the world." Bert was staring calmly at Jon at this point, while reaching for his glass of Coke.

"Since when?"

"Since the epistemological turn of the Enlightenment" replied back Bert.

"What is that?!" echoed both Ellie and Mary.

To have an epistemological turn during the enlightenment implies there were two different ways of knowing, two different ways of finding truth. We have a turn from one way of knowing to another that starts during the enlightenment period (1637 to 1800s, depending on author or source), and finished before the modern era. It was a switch in Western Culture from one epistemology to another. But to understand the switch and possibly why it happened we must first see the possible epistemology choices. Charles Kraft, Senior Professor of Anthropology and Intercultural Communication at Fuller Theological Seminary in California presented a chart that has been redrawn with modifications below for clarity; that describes the present worldview make-up of our world[14].

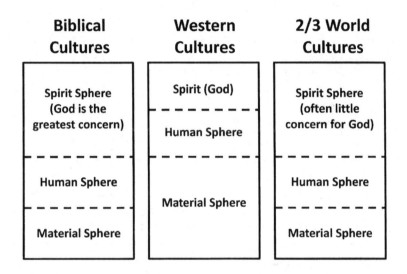

Biblical Cultures	Western Cultures	2/3 World Cultures
Spirit Sphere (God is the greatest concern)	Spirit (God)	Spirit Sphere (often little concern for God)
Human Sphere	Human Sphere	Human Sphere
Material Sphere	Material Sphere	Material Sphere

It divides existing and past worldviews into three categories: *Biblical Cultural, Western Culture, and 2/3 World Culture*. Each worldview is divided further into the spiritual sphere, the human sphere, and the material sphere. If we ignore the Biblical Cultures, we can split the world into the rationalistic West versus the more spiritually-minded East, Middle East, and Africa. Here in the West, control over material resources through technology, reason, and a more philosophically materialistic dominated epistemology reigns supreme. The rest of the world represented by the 2/3 World Culture places a much higher value on the spiritual sphere, and less faith on the material sphere. If we look at history we could argue that much of human history has been centered on a 2/3 World Culture worldview where the spiritual realm is dominated by pantheistic religions. Historically, the Greek and Roman worldviews introduced a sizable rational and deductive component into the Western culture but were essentially still pantheistic and 2/3 World Culture in their outlook. With the fall of the Roman Empire and the rise of the Christian medieval world we see the shift to the Biblical Culture worldview.

14. Kraft, "Worldview and Spiritual Power", 10.

It can be argued that the main historical architect of the Biblical Culture worldview going into the medieval period was Augustine. Augustine was an unusually bright product of his 2/3 World pantheistic Roman culture, and his conversion to Christianity was out of that culture. He was heavily trained in the rational, but skeptical Greek philosophy of Aristotle, Plato, Socrates, and their followers. These were called the peripatetic philosophers. Augustine used his peripatetic philosophical training in conjunction with his newly found Christian walk to construct a Biblical Culture worldview and defend his faith to the rest of the Roman world. His autobiography, *The Confessions of St. Augustine*[15], gives a stunning description of his thinking and experiences as he moved from pantheism to Christ. It was a spiritual and intellectual milestone that influenced the emerging medieval world greatly. His 2/3 World Culture worldview failed to deliver him from sexual sin, but his powerful new religion did. This new Biblical Culture worldview laid the foundations for the modern world and served Western Culture until the Enlightenment. Augustine's conversion to Christianity is a model of how a 2/3 World Culture view of life is rejected, and a new rational, but thoroughly spiritual worldview, is built. Augustine became disenchanted with his new age-like Manichaeism to reveal truth and deliver him from evil. His training in logic showed Manichaeism to be full of inaccuracies that initiates were supposed to swallow whole. After interviewing the charismatic leader of the Manicheans, Augustine declared.

> Nevertheless, I could remember many true things which the philosophers have said about this created world, and I could see the reason for what they said in calculation, in the order of time, and in the visible evidence of the stars. I compared their views with those of Manes, who, drawing on a rich vein of pure fantasy, has a lot to say on these subjects. I found in him no reason given for the solstices and the equinoxes or the eclipses of the sun and moon or anything else of this kind which I had learned in the books of secular philosophy. I was told to believe in these views of Manes; but they did not correspond with what had been established by mathematics and my own eyesight: in fact they were widely divergent.[16]

But the skeptical philosophy that delivered Augustine from Manichaeism could not deliver him sexual addictions. He soon realized that not every truth can be established by critical reasoning alone. Credible witnesses play a huge role in helping us determine what is true. The fact that he was

15. Augustine, Confessions.

16. Augustine, Confessions, book 5, chapter 5, 95–96.

born of his Mother and Father could only be firmly established by believing in the testimony of credible witnesses. But what really pushed Augustine over the edge of epistemological change; was deliverance from sexual lust, when he placed his faith and belief in Jesus Christ.

> Suddenly a voice reaches my ears from a nearby house. It is the voice of a boy or a girl . . . and in a kind of singsong the words are constantly repeated: "Take it and read it. Take it and read it." At once my face changed, and I began to think carefully of whether the singing of words like these came into any kind of game which children play, and I could not remember that I had ever heard anything like it before. I checked the force of my tears and rose to my feet, being quite certain that I must interpret this as a divine command to me to open the book and read the first passage which I should come upon . . . so I went eagerly back to the place where Alypius was sitting, since it was there that I had left the book of the Apostle when I rose to my feet. I snatched up the book, opened it, and read in the silence the passage upon which my eyes first fell: "Not in rioting and drunkenness, not in chambering and wantonness, not in strife and envying: but put ye on the Lord Jesus Christ, and make not provision for the flesh in concupiscence." I had no wish to read further; there was no need to. For immediately I had reached the end of this sentence it was as though my heart was filled with a light of confidence and all the shadows of my doubt were swept away.[17]

Long and McMurray[18] argue that Augustine built a Biblical world view with an epistemology as follows:

1. *Knowledge is power.*

2. *There are three possible sources of knowledge:*

   a. The rational knowledge attained by skepticism (later, the foundation for modern science)

   b. The occult, Gnostic knowledge of neo-pagan religion (the foundation of today's new age beliefs).

   c. The word of God, openly proclaimed in Jesus, and the Bible.

3. *The power most likely to tame our self-destructive, savage nature comes from the third source.*

---

17. Augustine, Confessions, book 8, chapter 12, 182–83.

18. Long and McMurray, The Collapse of the Brass Heavens, 71.

This epistemology served the western world for over one thousand years. But during the Enlightenment, the modern Western Culture worldview was started and the Biblical Culture worldview was rejected in ever increasing measure. Here at the Enlightenment the epistemological turn took place, and its main architect was Rene Descartes.

Descartes was a brilliant scholar and Catholic who wanted to prove the existence of God and the soul to nonbelievers; but only using experience and deductive reasoning. He rejected other forms of knowledge and restricted his search for truth to these two.

> By a method I mean certain and simple rules, such that if a man observe them accurately he shall never assume what is false as true, and will never spend his mental efforts to no purpose, but will always gradually increase his knowledge and so arrive at a true understanding of all that does not surpass his powers.[19]

Descartes restricted his epistemology to what could only be apprehended and controlled by the human mind. Long and McMurray have argued that it was the start of the abandonment of the Augustinian worldview. Descartes was criticized by people of his day as high in stature as the scientists Galileo, and Blaise Pascal. Pascal even wrote.

> I cannot forgive Descartes. In all his philosophy he would have been quite willing to dispense with God. But he could not help granting him a flick of the forefinger to start the world in motion; beyond this, he has no further need of God.[20]

From Isaac Newton to others, scholars and scientists, some Christian, some not; they gradually descended on Descartes philosophical works and built a new worldview that has become Western Culture or the modern worldview. Western culture possesses many very good things: the stunning successes of science and technology with commensurate good wrought in medicine, understanding, and the alleviation of human misery. But did we have to throw out God, the divine, the supernatural to achieve these things? Did we really achieve all of this good by pure reason, deductive, inductive, and empirical? I say no, we did not. We in the West have fooled ourselves into believing that the pure power of our minds did this: that Descartes' method expanded, made mechanical and material created all our good Western advances. Hopefully the previous discussion about the ways science really works has severely dented such an assumption. Science restricted to Descartes' vision is impotent, unable to really deliver all that it promises.

19. Descartes, Rene, "Rules for the Direction of the Mind", 74.
20. Pascal, Pensees, 77.

Descartes may have given the empirical sciences a starting boost, but they flourished in spite of their Descartesian base. Arthur Koestler argues in his book *Act of Creation,* that time and time again science made its most significant advances apart from logical positivism; Descartes' epistemology.

> Let us leave the borderlands of pathology. Nobody could have been further removed from it than the mild, sober, and saintly Einstein. Yet, we find in him the same distrust of conscious conceptual thought, and the same reliance on visual imagery.[21]

One of the most negative effects of a Descartesian epistemology, (or using our previous jargon, methodological naturalism), is that it leaves us with a universe that is philosophically flat and devoid of real hope. G. K. Chesterton slammed this worldview as follows.

> So these expanders of the universe had nothing to show us except more and more infinite corridors of space lit by ghastly suns and empty of what is divine . . . The idea of the mystical condition quite disappeared; one can neither have the firmness of keeping laws nor the fun of breaking them. The largeness of this universe had nothing of that freshness and airy outbreak which we have praised in the universe of the poet. This modern universe is literally an empire; that is, it is vast, but it is not free. One went into larger and larger windowless rooms, rooms big with Babylonian perspective; but one never found the smallest window or a whisper of outer air.[22]

So this is where Western Culture has gone. However in recent decades the problems of logical positivism and its descendent materialism have grown significantly. Western Culture is realizing that truth can be difficult to find, and science offers no sure route to absolute truth. Our culture is turning to the philosophical trends of Paul Feyerabend and his predecessors who argued that, "What remains are aesthetic judgments, judgments of taste, metaphysical prejudices, religious desires; in short what remains are our subjective wishes." Accordingly, there is no truth-finding method; we should follow what we want. But science cannot work, or function in such an environment for long. As Polanyi articulated so carefully, the search for truth is what drives every scientist's creativity. *It is faith in search of understanding; that here is the real and the true.* Perhaps this is why science is quite susceptible to darts thrown at its reason to exist. Both spiritually and pragmatically, science has not followed relativistic philosophy, but instead

21. Koestler, Act of Creation, 171.
22. Chesterton, "The Ethics of Elfland", 266–267.

retreated to a tempered sort of logical positivism. Here it can preach a pragmatic materialism, ignoring science's moral and spiritual components; but still use them behind the closed doors of its mind; lest anyone find out.

Science is split. It operates unconsciously under a moral and spiritual banner it cannot justify or rationally explain without appealing to God. But in the public square and its public voice, all is well and methodological naturalism is the secret to its modern success. What are the results of such a split in the mind of science? It acts as a filter that screens out and reinterprets data that reminds science of its moral and spiritual roots. Data is data, but it doesn't become science until it is sorted and interpreted. The devil lies in the details, *the interpretation*. Has modern science missed some of the most significant discoveries of the twentieth century because it could not handle the implications? I belong to the community of science; a field that I love. It is with regret I argue, yes it has. But that is a subject for our forthcoming chapters.

4

In the Beginning

A VERY, VERY LONG time ago there was nothing. At least we think it was a long time ago. This nothingness was not even nothing. If you were put into the middle of it (which would make it not nothing anymore); you could not go anywhere. This is because there was no place to go. You could not go left or right, up or down, forward or backward, because these simply did not exist. Even worse, you could not even be there and get older, time did not exist either. The dimensions of space and the dimension of time in which we travel execrably forward; always growing older, never younger; time wasn't there either. It is nothingness, that the human mind cannot wrap itself around; because we humans have no real analog or context with which to compare it. We best visualize this nothingness in little equations written on a piece of paper or some computer screen somewhere. But that is a poor excuse for the actual experience of nothing. But this is a start at least.

Into this nothingness came something. It was infinitely dense, infinitely hot, and full of energy; an energy, temperature, and density beyond our comprehension. But it appears to have come in this way, and it was our universe. Scientists call it the great singularity, because that is what you get when you divide by zero or multiply by infinity. A more popular name for this is the creation event; or the start of the big bang that gave rise to our universe. A few scientists say the great singularity has a cause beyond us and our universe. Some identify this cause as God. Other scientists and astronomers say it resulted because of a quantum fluctuation in the fabric of the larger set of multi-verses that randomly pop into and out of existence in the eternity of quantum fluctuation. These universe-sized quantum fluctuations can bring every possible universe into being of which the mind of man can conceive. This idea is given currency because evidence exists for subatomic particles popping into and out of existence of our universe's vacuum. But there is great difficulty with this idea; the vacuum of our universe is not empty, it has length and time, and some argue that a sea of sub-atomic particles waiting to be born resides inside it. Furthermore, there is no data or evidence at all to suggest that atomic sized phenomena can be applied to something as big as the universe. So what is this universe creating quantum fluctuation? It is a metaphysical idea that can be brought to the great singularity as a cause. It is fundamentally a philosophical idea rather than an evidenced-based scientific concept.

However another metaphysical idea can be brought to the great singularity with equal if not greater philosophical force. Something caused the great singularity, and that cause was God or something like God. There are two basic ideas wrapped up in these competing metaphysical strands, and they go as far back as the ancient Greeks. One: something has always existed and we and the universe somehow came from it. Two: God is eternal and created the universe and us. Take your choice, either God is eternal and always existing, or something else is; be it a place to have a quantum fluctuation or hold the many resulting universes or whatever. The big nagging question is why would we want to choose something other than God? Do we have a good reason to do so? Do we have a good scientific reason? The evidence for the quantum fluctuation model is severely lacking. Considering our previous discussions on how hard it is to separate our metaphysical assumptions from our science, it questions what legitimate philosophical grounds do we have for not considering God as a legitimate hypothesis for how the universe came into being, and why shouldn't the God hypothesis be given equal weight in our investigations.

What happened after the great singularity? From an infinitely small point it expanded and cooled, and expanded and cooled until it reached the size it has now. Most astronomers believe it took about 14.5 billion years for this to happen. And it continues to expand and cool. Many have likened this process to an explosion, a very controlled and balanced one however. It is so balanced that the term explosion does not do it justice. Those scientists who disliked the great singularity and resulting expansion sneeringly called it the Big Bang, the name has stuck. But what gave rise to the idea the universe was expanding from an initial creation event in the first place?

In 1914 Vesto Slipher reported to the American Astronomical Society, that he had observed some nebulae that were receding from the earth at a very fast rate. At that time in astronomy nebula were unknown astronomical objects; that were not stars but something else. In that meeting was a bright astronomy graduate student named Edwin Hubble. Edwin grasped that something major was afoot and determined to direct his research to finding out what it was. The resulting story is the stuff of legends. Edwin spent hours at the best telescopes of his day photographing these nebulae and collecting their spectra with the newly developed astronomical spectrographs that were attached to these telescopes. As a prism takes light from the sun and separates it into the various colors of the rainbow, Hubble collected the light from these nebulae and separated them into their respective rainbows or colors. However these rainbows had dark gaps and bright bands in them. The position of these dark gaps and bright bands (technically these are called absorption spectra or emission spectra imposed on a continuum background) were specific for each nebulae or star that was examined. You can use them to identify many of the elements (mainly hydrogen and helium) in these stellar objects. The positions of the dark bands and bright bands in the rainbow spectrum of these nebulae were well known. What Slipher had found was that the known positions of these bands in the rainbow spectrum had shifted their position, and shifted mainly to the red color end of the rainbow spectrum. What Hubble discovered was that the nebulae were not fuzzy gaseous object but separate huge galaxies with trillions of stars in them. Furthermore, these galaxy spectrums almost always had huge shifts in the known dark and light bands in the rainbow spectrum to the red side of the spectrum. This could only be explained if the galaxy was moving away from us if the shift was to the red side of the spectrum, and toward us the observers if the spectrum shifts were to the blue end. This was unusual and quite unexpected because the universe was considered as eternal by most scientists; and the stars were not supposed to be moving much at all. What Hubble had found was disturbing. These galaxies were moving away from us and doing it very, very fast. Why the fuss over every galaxy moving away from us? First, more information: Edwin Hubble also measured the distance from the earth to these galaxies. The distances are staggering, millions of light years (light year is the distance light travels in one year) away from us. The further a galaxy is away from us the faster it is moving away from us. Next; every galaxy was moving away from not only the earth but from each other as well. Every galaxy is moving away from everybody else. In short, the universe is expanding and the stars and galaxies in it are moving away from each other as it expands and the ones at the outmost edges appear to be moving the fastest. Now play God and video this whole process as time develops. Then run the video recording backwards and watch what happens. The universe keeps shrinking and eventually it gets so small that it disappears and ceases to be—the great singularity. An infinitely large density divided by an infinitely small volume gives an infinitely large energy density and infinitely high temperature—the creation event. Conclusion, the universe wasn't always here, it had a beginning. Something from nothing demands a cause, and a cause sufficient for the effect of the universe is generally called God. So much for an eternal universe, maybe something else had to be eternal; like God?

Dr. Hanson was an impressive person. He was not huge, but he wasn't small. His presence sometimes seemed to take over the room and wash over you like a tidal wave. Many a student had been reduced to stammering when queried by Dr. Hanson. Many loved him; others wanted to run for their lives. But lecture was never dull when Hanson took over. Hanson knew the Hebrew and Greek languages and ancient cultures like few others. But Hebrew was his specialty and no quarter was given to those who disagreed.

"When you study an ancient culture you must think as they thought, breathe as they breathed; see as they saw. No room for your quaint western culture that dices up the world into logical propositions that fit so nicely. Nature is not a gentle field of grass with ecologists dancing across the surface, while research papers and movies drool out the latest scientific finding," stormed Dr. Hanson. "No! Nature is fierce; it can kill you in a moment's notice. Your neighbor is a foreigner who could at the slightest provocation grab his sword and kill you; kill you because of an incident you had no control over whatsoever. Kings were absolute, and extreme in their mercy and harshness. Perhaps the gods that ruled these uncontrollable forces of nature and people were the same."

Pausing a bit, Dr. Hanson continued, "Our earliest creation story, or creation myth as professional linguists like to call them, is the *Enuma Elish*[1]. It is the Babylonian creation story involving their patron god of Babylon, Marduk. The title comes from the first two words which are translated as "When on High . . . ." The story comes from clay cuneiform tablets unearthed in the middle of the nineteenth century in the ruins of the palace of Ashurbanipal in Nineveh. The tablets date to 700 BC, through the story dates to 1200 to 1500 BC most likely. Liberal theologians argue that the Biblical Genesis creation account borrows from the *Enuma Elish* and is a later theological evolution from the Babylonian creation myth. Modern evangelical and conservative scholars argue that the accounts are just too different; but agree the Genesis writer freely used metaphors and symbolism drawn from their common Mesopotamian cultural pool.

Staring at the first row to accentuate his point, "I will suggest another possibility. The Genesis writer wanted to contradict the Babylonian creation

---

1. Enuma Elish.

story and make some very definite points. Our writer wanted to correct the theological errors of the common Mesopotamian creation account and set the record straight; and make some very definite theological statements in the process. Hence, the common metaphors make sense, but the purposes and theology of our writer are utterly different. To see this, let me explain the *Enuma Elish* myth a bit."

Mary was becoming interested in the Bible. She had been raised in the church but was unsure as to what she really believed. Dr. Hanson's class seemed quite interesting. Perhaps it would answer some questions she had about the Biblical creation story. She wasn't surprised to see Ellie in the class, but why in the world did Jon want to take it?

"So Mary, do you have any ideas on what points the Genesis writer wanted to make?"

Jolted from her thoughts, she sputtered out, "I don't know yet, I have to hear the *Enuma Elish* first."

With a sly grin Dr. Hanson, spoke, "It was assigned reading that you should have read before class, but fair enough, let's talk about it."

"Dang it!" thought a red-faced Mary, "Why doesn't he give us a chance to catch up with him."

"Let's set the stage for our story. Apsu is the god of fresh water and thus male fertility. Tiamat is his wife and the goddess of the sea, and likewise chaos and threat. They have kids who are gods, Anshar and Kishar, who in turn have a son god who in turn bears Ea and others. These sons of the gods make such a racket and fuss that Apsu and Tiamat cannot sleep or rest. Apsu decides to kill them all so that he can get some peace. Ea the son of the sky god hears of the plan and beats Apsu to the murder by killing him first. Then he and his wife Damkina establish their dwelling place above the body of the dead Apsu and give birth to Marduk the patron god of Babylon and the god of Spring.

Mary was busy scribbling the names down as fast as she could, "This prefigures Genesis? You have got to be kidding me."

"Tiamat, who wasn't sleeping well either, is obviously enraged that her lover and husband was killed by one of her descendent gods. She vows revenge, creates eleven monster creatures to help her carry out her revenge. She takes on a new husband god named Kingu, and puts him in charge of her army of monsters to carry out her plan. Tiamet is the goddess of winter, or disorder and chaos in the world, and plans to unleash that chaos on the other sons of the gods in their destruction. Ea learns of her plans and tries to talk her out of it, but to no avail. As the other gods become aware of her plan they try to drum up courage to face her but fail and fall back in fear. But at last brave Marduk steps up and agrees to battle Tiamat; *Tiamat, a woman who flies at you with weapons.* That was a quotation from the *Enuma Elish*, not big on woman's rights are they?" At this point Dr. Hanson was pointing his finger at the second row of busily writing girls. A guy in the back snickered, but a glare from Dr. Hanson silenced him immediately.

Continuing the drama, he raised his voice a bit and said, "Marduk is appointed king of all the gods and commissioned to fight Tiamat. He rides his chariot of clouds and assembles the four winds to help him fight Tiamat. As the battle ensues Tiamat opens her mouth to swallow him whole, Markduk entangles her with his net, and then inflates her with the Evil Wind until she is incapacitated. An arrow through her heart kills her; and her army of gods and monsters flee in fear. Marduk captures them all including her consort husband Kingu, and throws them into prison. He smashes Tiamat's head with a club and divides her corpse in half. One half is used to create the earth, and the other half is used to create the sky. Marduk creates dwelling places for his allies and thus creates the Babylonian calendar using their constellations in the night sky. But the gods need someone to serve them in the heavenly abode, to save them from menial labor; so Marduk the king of the gods decides to create mankind, the black-headed ones who can perform menial tasks for the gods. He inquires who incited the Tiamat rebellion and finds that it is Kingu, Tiamat's second husband. Ea kills Kingu for Marduk and uses Kingu's blood to fashion mankind. And then the creation myth finishes off with instructions to black-headed man to honor and serve Marduk."

Mary was fully awake, amazed at the banality of this creation tale. "These people just pulled their creation story from the soap opera politics of their kings and queens. Is this the best that they could do?"

"Okay class, what do you see?" exclaimed Hanson. "Politics and squabbling," spoke out Jacques on the third row. "They based their gods and creation story on their own experiences and sordid tales."

"Good"

"The gods are not all powerful but have human characteristics, i.e. they are just super human versions of us."

"Very interesting!"

Mary raised her hand, "The creation of the earth seems to be a sort of accident or after-thought."

"Very good! Did you catch that class? Creation doesn't seem to have much purpose in it, it just happened, almost accidental," Hanson glanced at Mary and said, "Now what about the creation of man?"

"Marduk or Ea used Kingu's blood to make man at the very end of the story," sang Alan at the back of the class.

"Correct," intoned Dr. Hanson. "Man is an after-thought, an accident if you will. There is no real purpose in creating him other than to serve the gods in menial tasks. He was made out of the blood and guts of dead Kingu, a dead, vanquished worthless deity. Now think very carefully, is this similar to any modern creation myth/story in our culture?

From the right side of the room, Jon spoke up, "The modern meta-physical story of evolution in western culture; forces and particles to man. Accident brought it all together."

"Very interesting; very, very interesting," said Dr. Hanson. "Like the writer in Ecclesiastes said, "there is nothing new under the sun." There really isn't; is there?"

Mary thought to herself, "This *is* interesting!"

The next day Dr. Hanson strode into the class with Bibles and text-books under arms. "Now class, let's examine the Christian creation myth, today." He plopped the texts down on the front desk and rapidly started up his laptop and computer projector. "If you buy that the long standing traditional author of Genesis is Moses then sometime around 1446 to 1406 BC, Moses penned the following to start off Genesis." The projector stirred to life, flickered, and then flashed the following on the screen in very large letters for all the class to see.

In the beginning God created the heavens and the earth. Genesis 1:1 NIV

"By the way I tend to buy the traditional authorship of Moses, but even if I am wrong, it changes very little of what I have to say." Dr. Hanson looked over his glasses at the class as if he actually expected a sort of scholarly challenge to be issued. Seeing none he resumed. "What do you see in this little verse here class? What first leaps out at you?"

From the back a student timidly ventured, "In the beginning, it starts with in the beginning"

"Yes, yes it starts with in the beginning, but what leaps out at you, grabs you by the throat, argh, argh." At this point Dr. Hanson was grapping his throat and pretending to be strangled to emphasize the point.

"Ah, God created the heavens and the earth" slipped out another student quickly.

"Yes, yes, GOD CREATED, God created," exclaimed Dr. Hanson. "Marduk did not create, Apu did not create, the gods of the heavens did not create, but GOD created. Right in the very first sentence Moses is making the most definite in your face monotheistic statement he could make. Moses' culture was Mesopotamian, Babylonian. Gods were a dime a dozen and understood by all to have made the heavens and the earth. And our Moses has the gall to say God created it all. Put yourself in his shoes, his entire culture believed otherwise and out he comes with GOD CREATED. Notice what he created out of—Nothing. Before creation there was only God, not gods and something, but God alone, and then He created the heavens and the earth from nothing. Did Marduk create the heavens and earth from nothing? I think not; and Tiamat's corpse agrees with me! Theologians and philosophers have a name for creation out of nothing, *creation ex nihilo*.

It is fascinating that in this century the findings of modern astronomy can very easily, if not most rationally, support the idea of creation ex nihilo. Not everyone agrees obviously, but the case can be made and with very good cause. The Genesis text appears to stand alone in the ancient world in asserting that the world was made out of nothing. How could the Hebrew civilization of 3500 years ago get it right when every other civilization of their time, and thousands of years after, got it wrong? What was the basic view of the world in the Mesopotamian milieu in which the Hebrews lived?

From ancient writings, excavated clay tablets and other artifacts we now have a better picture of what the Babylonian/Mesopotamian universe looked like. The world was flat and circular, surrounded by water on all sides. If you were to go to Palestine or the Arabian Peninsula and head west you hit the Mediterranean Sea; go east or south, the Persian Gulf, the Red sea, or the Indian Ocean. Thus the world was flat like a pancake with water all around the edges. Under the world was water, the "waters below" and over the top of the world was stretched the sky like a huge canopy or dome, from which the stars, sun, moon, and planets were hung. The "firmament" which stood at the edges of the world supported this dome of the starry heavens and daytime sky. Above this dome was a great deal of water, the "waters above," and sometimes it leaked through the canopy of the sky, and you got rain. The sun left its "house" underneath the earth, and traversed the daytime sky and then returned to its night time abode beneath the world to return the next day.

"Okay class, let's return back to the Genesis narrative," Dr. Hanson has just finished describing the Mesopotamian worldview in exquisite detail, "Matt, why don't you read until I tell you to stop." Matt being caught unawares by this request fumbled for a Bible and finding he had none nervously responded, "I don't have one, sir." Dr. Hanson glared at him with a look teetering between disgust and amusement replied, "A class on Hebrew and Genesis and you don't have a Bible? Does anybody here have a Bible that they can read from?" Seeing no immediate takers, Dr. Hanson moved the mouse to activate his Apple computer (no self-respecting linguist would ever be without his Apple PC), clicked to his 50 version Bible program. "The NIV gets some of the Hebrew wrong in my opinion so let's try the NKJV or NAS. Verse two, "The earth was formless or void and darkness was over the surface of the deep, and the Spirit of God was moving over the surface of the waters." The Hebrew word here suggests brooding," he intoned. "Then God said, "Let there be light"; and there was light. And God saw that the light was good and separated the light from the darkness. And God called the light day and the darkness he called night. And there evening and there was morning one day.

"Whoa!" thought Jon, "Surely this isn't referring to the end of the dark phase of the universe when it was so hot that photons and other subatomic particles could not yet exist, then as the universe expanded and cooled, photons and the possibility of light suddenly appeared?"

His thoughts were interrupted, when Dr. Hanson boomed out, "And there was evening and there was morning, one day. Make note of this. Ancient Hebrew had no chapters, paragraphs, punctuation, or other text based markers to indicate the end of an idea or concept. So how did they do it? They marked it by repeated phrases or textual markers that were fairly obvious to the reader. Perhaps like non-literate oral cultures of our modern era, who used poetic forms to structure their verses, and epic poetry to make it easier to memorize and organize in their heads; instead of on a piece of paper like we literate moderns do. Reading on, Dr. Hanson stated, "Let there be an expanse in the midst of the waters, and let it separate the waters from the waters, yada, yada, yada. And God called the expanse heaven, and . . . okay class."

A third of the class replied, "And there was evening and there was morning a second day."

Going on, "Then God said, "Let the waters be gathered together into one place, and let the dry land appear," yada, yada, yada. And the earth brought forth vegetation, plants yielding see after their kind, and trees bearing fruit, with seed in them after their kind; and God saw that it was good." Dr. Hanson looked up at the class waiting expectantly for the response.

One half of the class replied, "And there was evening and there was morning a third day."

"Okay class, what do we have here? First God created light and darkness and the space that contained it. Next God created the heavens that separated the waters above, from the water below. Then God separated the waters below so that the dry land could appear. Three different spaces; now what is God going to do with them? Let's read on. In verse 14 God says, "Let there be lights in the expanse of the heavens to separate the day from the night," etc., etc., etc., "And God made the two great lights, the greater light to govern the day, and the lesser light to govern the night; He made the stars also, "yada, yada" and God saw that it was good."

Again the class replied, "And there was evening and there was morning, a fourth day."

"What has happened here class?" Dr. Hanson looked up quizzically.

Alicia spoke out, "God has filled the expanse containing the light and darkness with the sun, moon, and stars."

"Right!" exclaimed the professor. "Now let's continue on. Verse 20, "Let the waters teem with swarms of living creatures, and let birds fly above the earth in the open expanse of the heavens. And God created the great sea monsters, and every living creature that moves, which the waters swarmed after their own kind; and every winged bird after its kind; and God saw that it was good." Etc., etc., "and God saw that it was good." Hanson looked up waiting again.

"And there was evening and there was morning, a fifth day."

"And . . . ?" Dr. Hanson asked.

"God filled the expanse of the heavens and the waters below created on day two with birds and sea creatures," a student in the rear loudly proclaimed.

"Then God said, "Let the earth bring forth living creatures after their kind; cattle and creeping things and beasts of the earth after their kind;" " Hanson paused a bit and resumed. "Skipping ahead a few verses we find, "Then God said, "Let Us make man in Our image, according to our likeness; and let them rule over the fish of the sea and over the birds of the sky and over cattle and over all the earth, and over every creeping thing that creeps on the earth. And God created man in His own image, in the image of God He created him; male and female He created them; and God blessed them, and God said to them, "Be fruitful and multiply, fill the earth, and subdue it, rule over the fish of the sea and over the birds of the sky, and over every living thing that moves on the earth." " He continued on to verse 31 reading, "And God saw all that He had made, and behold, it was very good. And there was evening and there was morning, the sixth day."

"So class, again we have the space created on the third day, the dry land, filled with creeping creatures and lastly with man himself. Do you see the literary form or structure the author was using? You create a space and then fill it. Day one creates the space for light and dark and then day four fills it with the sun, moon and stars; day two creates the waters below (the seas), and the heavens and fills it with sea creatures and flying birds, in day five; and day three creates the dry land and then fills it in day six with land animals and finally man himself. See the structure? With a quick step he strode to the dry erase board and quickly drew:

God creates ex nihilo

Day 1—Light and darkness	Day 4—Fills it with sun, moon, and stars
Day 2—Sky and waters below (the seas)	Day 5—Fills it with birds and fish
Day 3—Dry land	Day 6—Fills it with land creatures, and man

Day 7—God rests

"Now bear in mind the Hebrew mind and worldview of Moses day was primarily an oral culture, they had writing, but writing materials were rare and very expensive. Extraordinary care was taken in reproducing manuscripts especially when it was considered the Word of God. But they did not think in linear, concise, precise, scientific terms like we do today. To try to impose that type of thinking on their culture is to do a great insult to their thought, culture, and wisdom." Dr. Hanson always kept a bottle of water at hand during lectures and proceeded to down a huge swig before proceeding. "Their culture valued wisdom, not necessarily knowledge for knowledge's sake, but knowledge that did something for your life, that allowed you to live better in a harsh and extreme world. Poetic structure, ease of memorization, profundity would have been much more important to them then some alleged scientific order, which would have been unknown to them anyway."

"Dr. Hanson," asked Ellie, "Do you believe, The Genesis account was inspired, written by God, and accurately reflects the way the world was made?

"Yes."

"Then why is it, what we have just gone over is out of order with what science has taught us about the formation of the universe and world. Who is right, Genesis or science? Does the Genesis record contain errors in your opinion?"

Hanson paused a bit before replying, sipping from his water to gather time, "Ellie, you are not listening, think before you impose a modern perspective on the text, you insult the author and what he is trying to say if you do. Just because the writer of Genesis doesn't think scientifically, whatever the heck that really is anyway, doesn't mean he is stupid, incorrect or even inaccurate. Moses or whoever was the original writer, was brilliant, and heard from God dead accurately in my opinion. It is just the way he in that culture and era is going to express it is going to be very different than how we would express it in our day. John Calvin the great Protestant reformer of the 1600s expresses it this way. God has to accommodate his knowledge and ways to our very limited human understanding. How God does that in different eras of human understanding is going to be different because we humans are at different levels of understanding. Suppose God told Moses that the world was not flat but round; how would the culture of Moses' day take what he said seriously, when everybody then with any real intelligence and learning knew the world was flat. But yet how do you communicate the reality of how the world was formed to such a culture so that 2000 to 3000 years later it still communicates the truth of how the world came to be?" Hanson looked around and raised his hand to emphasize the point. "Think deeply on this, what makes you think that God has stopped accommodating to our present scientific culture? May I suggest that we are still far too stupid to really understand how the world came to be. Who do we think we are; to believe that because of our "superior" scientific knowledge, God has to accommodate to us less than he does to Moses and his day? Maybe he has to more because we have become so dense in our present day understanding of wisdom. You think I jest, look at the text carefully again, what does it say, class." He peered intently in a sweep of his steely eyes at the class looking for signs of intelligent discourse.

Ellie trying to rescue herself sputtered out, "There is purpose and intentionality on every verse of the creation account, God saw that it was good, He made it that way, and He wanted to do it."

"Yes very good. This is in my opinion a direct in your face contradiction of the current Babylonian view of the world, a setting of the record straight, and a slap in the face against the multitudes of gods that everybody believed in. Moses is saying, "There is one God, there are no others, and this God alone is responsible for making the world, and he did it because he wanted to, and he made it good. It was no accident." Now notice the ending of the seventh day in verses two and three of chapter two."

Jack chimed up, "There is no there was evening and there was morning the seventh day tag for that day."

Hanson put his hand to his chin rubbed it and said, "Things that make you go hmm!" He quickly turned to the other side of the class who were intently staring at the professor, "Hmm, why did our author do this? Was it an oversight, an accident, or was he trying to tell our readers something. Jack read Hebrew 4:4, 5 and Hebrews 4:10."

Jack read, "For He has thus said somewhere concerning the seventh day, "And God rested on the seventh day from all His work;" and again in this passage, "They shall not enter my rest." Verse 10, "For the one who has entered His rest has himself also rested from his works as God did from his."

"Okay class, what does this mean?"

"We are still in the seventh day," Jack replied.

"Yes!" Dr. Hanson exclaimed, "We are still in that seventh day when God is resting from his creative activity. Solves a lot of problems, Genesis as a literary motif has no conflicts with most of the findings of modern science, we have not sacrificed the truthfulness, accuracy, integrity of the text, and it follows logically from an ancient cultural sensitive reading of the text"

Ellie raised her hand, she was disturbed, "But Dr. Hanson, aren't you reducing Genesis to a text that can mean whatever you want it to mean and thus it loses it truthfulness!"

"How so?"

"Well if Genesis one can be read that way then the crossing of the Red Sea and all of Genesis can be read like that."

Seeing a bit of panic in Ellie's voice, Hanson paused a bit before replying, "Ellie, three points; one—I never said this was the final answer. Two—in education you need to look at all views whether you agree with them or not, if you want to be educated. Three—no you cannot read the majority of Genesis like I was reading Genesis one. Most of Genesis, the crossing of the Red Sea, Abraham's journey's, etc., is written as historical accounts, the Hebrew screams it, and thus must be read that way. The Hebrew text clearly says that the waters were piled up on each side like a wall, when Moses led them through the Red Sea. The text indicates a supernatural miracle, not coincidental winds blowing the waters back into a nice thin sheet to keep the non-miracle theologians happy. The Hebrew is very clear; piled up, and it is not in a structured poetic format like Genesis one is. Personally I stand by the miraculous reading of the Red Sea crossing, but that is my opinion. That is one of the beauties of the literature based approach to the Old Testament. You know when you can and generally cannot apply a poetic format; you cannot do it willy-nilly as you please to fit your personal theology. It treats the authors as intelligent people who had very specific reasons for writing and communicating as they did and it does not in my opinion compromise the text with respect to its inspired nature and truthfulness. We call this reading of Genesis the framework hypothesis within a literature based hermeneutical approach to the Scriptures. Does it solve all of our problems or represent the only way to read Genesis?"

"No!" replied Ellie, "Why can't we read it in the clear sense of what it says, seven 24 hour literal days! Jesus and Moses referred to the creation account as if they were seven literal days."

"Ellie, you are right, that is a problem for the framework hypothesis, but I do challenge your common sense reading being a literal 24 hour day. For example look at chapter two of Genesis. First where is the real start of chapter two in the text? Remember verse numbers and chapters were added sometime around 1000 AD by monks who were not the best theologians, and some of their choices for chapter headings make no sense at all. Look at Genesis 2:4. "This is the account of the heavens and the earth when they were created, in the day the Lord God made earth and heaven."

Hanson paused again and put his hand on his chin and said "Things that make you go, hmmm."

Jack echoed back, "Moses is trying to tell us he has finished the first account and now is starting on his second account; the second chapter ought to start here."

"Yes, very good, very good. Now look at the order of creation in chapter two as compared to chapter one, are they the same?" No one answered back, they were too busy reading and digesting this thought. "No! It is different. Is our author stupid? Unable to count or remember what he wrote just a page earlier."

No one answered though Jon contemplated a reply; but decided against it.

Hanson lifted his hands as if in futility, "No! He is not stupid; he knew exactly what he was doing. He is trying to tell us that the order is not important to his account. I argue he did this intentionally. Give our author his due; he knew what he was doing; whether through the inspiration of the Holy Spirit, directly or indirectly, I cannot say. Stop treating our author as an ignorant ancient sheep herder. He was a giant of his age, and of our own also."

Class was almost done. The professor finished up by glancing at his watch and announcing in his booming voice, "Class, our time is almost up. Is this the final say in reading Genesis that you will run into in western culture? Absolutely not! There are old earth day age type readings, young earth readings, gap theory readings, remember the old Scofield Bible. There are theistic evolutionary readings, and other evolutionary creationism type viewpoints. And even some combinations of old and young earth readings. We will talk about these in latter classes. Class dismissed!"

As Ellie gathered her books and wandered out the class, Jon strode up beside her and quietly smirked, "Poor Moses can't keep track of his order and days, can he?"

Ellie with fire in her eyes shot back, "Creation ex nihilo, big bang!"

"Oh be quiet, you silly pretty girl" he replied, and then strode off down the hallway, never glancing back to see the needles darting out of her eyes.

The astronomical evidence for a beginning to the universe is very disturbing to a materialist who posits an eternal something other than God from which to form the universe. But more disturbing is that the universe appears to be adjusted or "fine-tuned" for life and possibly for mankind uniquely. This adjustment is like winning the lottery where the stakes are beyond belief and the odds against winning it so high that to win makes you suspect the whole lottery has been rigged. In this case the rigging would have to come from a supernatural agency of immense powers (God) or some other cause so powerful and intelligent acting that we might as well call it God. How far down the structure of our universe, does this fine tuning go? I will argue it goes all the way down, and into areas we have just begun to research and understand.

Michael Behe in *The Edge of Evolution*[2] provides a chart illustrating the depth of fine tuning in the universe. Here I have redrawn a version that makes a good starting point for our discussion.

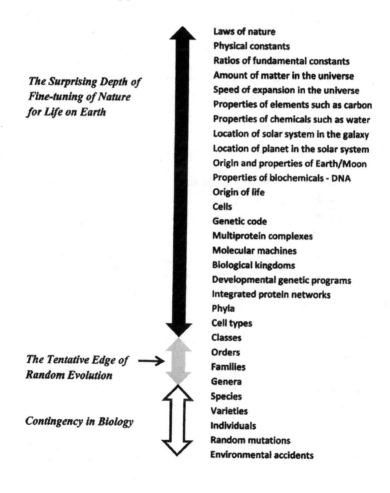

The Surprising Depth of Fine-tuning of Nature for Life on Earth

The Tentative Edge of Random Evolution

Contingency in Biology

Laws of nature
Physical constants
Ratios of fundamental constants
Amount of matter in the universe
Speed of expansion in the universe
Properties of elements such as carbon
Properties of chemicals such as water
Location of solar system in the galaxy
Location of planet in the solar system
Origin and properties of Earth/Moon
Properties of biochemicals - DNA
Origin of life
Cells
Genetic code
Multiprotein complexes
Molecular machines
Biological kingdoms
Developmental genetic programs
Integrated protein networks
Phyla
Cell types
Classes
Orders
Families
Genera
Species
Varieties
Individuals
Random mutations
Environmental accidents

2. Behe, The Edge of Evolution, 218.

First let's digress into the meaning of "fine-tuning" and how we wish to use it here. Something is fine-tuned when a particular value is chosen out of a very wide range of values, and only that value or something very close to it will give you the desired result. In our case, that result will be "life," living things, the possibility of life, or a universe capable of supporting life. If the range of values we can choose is restricted by outside conditions then our value is not fine-tuned, just inevitable. If the range or width of values we can choose out of the available set of possible values is very large, then it doesn't matter much which value we choose, any old value or setting will work and give us "life." Imagine an older radio where you turn a knob to dial in a radio station by rotating until you reach the desired frequency and station. You have a lot of frequencies to choose from but only one will give you KFAQ Tulsa's, only clear channel talk radio station. If KFAQ is at every frequency except the very low or very high end of the AM band, then it doesn't matter likely where you set the frequency dial, just give it a spin and out will pop KFAQ. However if only 1170 kHz will do, then you must be very careful to set the dial right or no talk. Just giving it a spin will not do the job.

Question: who turns the dial for the universe? If the universe made itself or existed eternally then the universe is responsible for turning the dials either by some unknown feature or the laws that govern how the universe works. If God or a super intelligent agency did it, then He/they turned the dials or made the laws that turned the dials. If we have reason to believe the process that produced the fine-tuned feature of the universe, arose randomly, (scientists call these stochastic processes), then the likelihood of accidentally getting that fine-tuned value becomes important. Two things affect the possibility of getting a fine tuned value through a random or stochastic process. How many tries do you get in the time available to get it; and how likely is it that you will get it. The wider the range of possible values you have available, and the finer the exact value you must choose, then the less likely that you will ever get it. But if you have an infinite number to tries to get it, then eventually you will hit the magic value in your infinite number of tries. But if you do not have an infinite or incredible large number of tries to hit upon the right fine-tuned value, then you have good reason to suspect that the correct value was "selected" by an intelligent agency. You can argue that the universe was "designed" by an intelligent agency (God) with foresight and knowledge to select the correct values from among the many possibilities.

The measure of how much time and how many tries a random process has to randomly hit upon a given fine-tuned value can be measured by the *probabilistic resources* available to the stochastic (random) process. If the universe is eternal, then the probabilistic resources for any random process to achieve the fine-tuned value are infinite and it can happen; maybe. If the universe as we know it is not eternal, but finite: then the probabilistic resources are limited. If your probabilistic resources are not great enough to meet the probability demands of the fine tuning you are examining, then you have very good reason to infer an intelligent agent monkeying with the dials to get the right number. "Somebody" set the dial. That is, the universe was "designed."

What are some upper bounds or numbers we can use to set an upper limit for the probabilistic resources of the entire universe? If our random process demands more than is available, then we can safely declare the random process as impossible, and look to an intelligent cause for the event that happened. Current estimates of the age of the universe, if one accepts the ancient old earth/universe hypothesis, places the age of the universe at 14.7 billion years old[3], the age of the earth at 4.5 billion years old[4], the age of living things from the fossilized record at 3.5 billion years ago[5], the age complex multicellular creatures with exoskeletons and spinal cords appeared 525–530 million years ago[6]. From astronomy the number of atoms in our own Milky Galaxy[7] is about 10^{65} and the number of known elementary particles in the universe[8] is at about 10^{80}. From these estimates we can construct a set of probabilistic resources available to random processes to get the job done. If they require much more, then it is high reason to suspect an intelligent designer "rigged" the lottery.

Now is a good time to introduce the concept of combinatorial inflation. The mathematical field of combinatorics studies how a group of things can be combined or arranged and calculates the number of ways it can happen. If the number of ways a random process can happen requires more probabilistic resources than the universe provides, then the process cannot happen, at least not *randomly*. Below is a diagram of combination locks based on the combination lock diagrams shown in Stephen Meyer' book, *Darwin's Doubt*[9]. It illustrates the problem of combinatorial inflation. Imagine a bike lock with rotary dials that can be set. A three dial lock has 10 digits per dial. Thus there are 10^3 or 1000 possible combinations for this lock. Likewise for a five dial lock one would get 10^5 or 100,000 possible combinations. A ten dial lock has 10^{10} or 10 billion possible combinations. A hundred dial

3. Ross, Improbable Planet, 48. Author's note: the precise age of the universe is still uncertain to 0.2 - 0.4 billion years, so I chose the average of 14.7 for the two values of 14.5 or 14.9 billion years given by Ross and his selected references.

4. Schopf, Cradle of Life, 4.

5. Schopf, Cradle of Life, 100; Ross, Improbable Planet, 17.

6. Meyer, Darwin's Doubt, 72.

7. Meyer, Darwin's Doubt, 175.

8. Meyer, Darwin's Doubt, 175.

9. Meyer, Darwin's Doubt, 174.

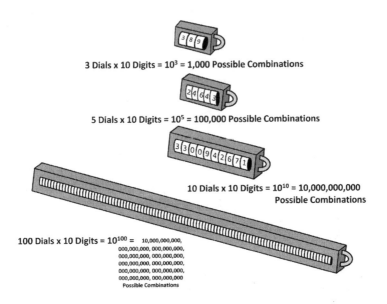

3 Dials x 10 Digits = 10^3 = 1,000 Possible Combinations

5 Dials x 10 Digits = 10^5 = 100,000 Possible Combinations

10 Dials x 10 Digits = 10^{10} = 10,000,000,000
Possible Combinations

100 Dials x 10 Digits = 10^{100} =   10,000,000,000,
000,000,000, 000,000,000,
000,000,000, 000,000,000,
000,000,000, 000,000,000,
000,000,000, 000,000,000,
000,000,000, 000,000,000
Possible Combinations

lock has 10^{100} combinations which exceeds the number of particles in the known universe by 10^{20}. How fast you can dial in a combination times the time you have to find the one given combination that opens the lock is the critical factor. If you can dial in a combination every 6 seconds and you have 2 hours to try to guess the lock combination of a three dial lock, you can go through 10 tries per minutes times 120 minutes to get 1200 tries. You will likely guess the combination of the three dial bike lock and open it. You have 1 in 1000 chances of guessing the right lock combination. That is you have a probability of guessing the right combination per try is 1/1000. But 2 hours (120 minutes) at 6 seconds per guess gives you the probabilistic resources necessary to open the lock. One thousand possible lock combinations are called the "combinatorial space" of your bike lock. The number of possible bike lock combinations you have to search over to find the bike lock combination is called the "combinatorial space" of the bike lock.

If we enlarge our bike lock to 5 dials but keep the time, and time per tries the same, then it is not likely we will find the correct bike lock combination. We have a 1 in 100,000 chance of getting the combination but only the probabilistic resources of 1200 guesses to pull it off in. Not likely, but maybe possible if we get really lucky. If we enlarge our bike lock to 100 dials the numbers go out of sight. We have a one in 10^{100} chances of dialing in the lock combination right on the first try; but only 1200 tries in our two hours to get it; an impossible scenario. Anybody who guessed the correct combination of a 100 dial bike lock under the 2 hour limit would be accused of having some fore-knowledge of what the correct combination was. The probabilistic resources available to them were not sufficient to guess the bike lock combination, they had to have known or had information about the correct combination. To say it was random would be irrational under these circumstances, or suggest that you understood little about how probability really works.

There is a subtle difference between our previous bike lock scenario and the random processes that often take place in the real world. In the above examples, each guessed combination is tried only once and then a new different combination is tried next. A truly random process doesn't have the knowledge to try a new different combination on every guess. It would be like spinning the dials randomly on each dial to arrive at the new guessed combination every 6 seconds. You would often repeat a previously guessed combination. Thus your search through the combinatorial space of your lock would take even longer, requiring even more probabilistic resources.

Notice how the probabilistic resources needed to randomly find a given combination expand exponentially with each addition of a new dial on the bike lock. This rapid inflation of the probabilistic resources needed, as the dials are added, is called *combinatorial inflation*[10]. Now change the dials to amino acids on a protein, base pairs in a DNA molecule, combination of possible fundamental constant values among those that produce life sustaining universes, among all possible combinations or ranges of values. You will quickly find yourself with the problem of combinatorial inflation versus the probabilistic resources necessary to find the values that give us a universe suitable for life.

The materialist needs a universe that can be assembled by blind chance or else a set of eternal laws that force the universe to come to be as it has such that it has life and beings like us. If we cannot find the principles or evidence that "forces" the universe to be as it is, then a materialist can posit random processes and lots of time to produce the fortunate happy living universe we inhabit. If the probabilistic resources of the universe are not sufficient to allow such random construction of the universe or living things within it, then alternate explanations must be pursued. The only alternate explanation that fits with our common experience is that living things and the universe were "designed" by an intelligent agency, (God, in my opinion). This common experience we will discuss latter.

The problem of combinatorial inflation versus the probabilistic resources necessary to construct a life giving scenario by random processes appears often in the scientific study of life. One could legitimately argue that

10. Meyer, Darwin's Doubt, 169. Meyer discusses this in depth in chapter 9.

the more we study and understand the mechanics of life, and the conditions necessary to sustain it, the worst the problem has become. This has not been lost on some scientists. The great astronomer, and shaken atheist, Fred Hoyle wrote in the November 1981 issue of Engineering and Science.

> A common sense interpretation of the facts suggest that a super intellect has monkeyed with physics, as well as with chemistry and biology, and that there are no blind forces worth speaking about in nature. The numbers one calculates from the facts seem to me so overwhelming as to put this conclusion almost beyond question.[11]

Here Hoyle was reacting to his discovery of the fine tuning of the nuclear resonance in the carbon atom that allows carbon and other heavier elements to be formed in the nuclear fusion of the stars. He was greatly amazed by this and other examples of fine tuning in the fundamental constants of the universe. Well how many fine tuning scenarios are there, that require probabilistic resources that our current conception of the universe cannot supply? Go back to Michael Behe's chart of the depth of fine tuning in the universe and life. Every label by the vertical black bar has many of them; some that are inferred by simple understanding of the processes involved, and others where actual estimated numbers have been calculated. Well what are these fine-tuned cosmic constants? That depends on how you look at them and what you are going to do with them. If you want to list constants whose values must be precisely so for conditions in the universe so that life exists one can find a list of over ninety such values[12] such as astronomer Hugh Ross tabulates in his books and website.

1. Strong nuclear force constant

2. Weak nuclear force constant

3. Gravitational force constant

4. Electromagnetic force constant

5. Ratio of electromagnetic force constant to gravitational force constant

6. Ratio of proton to electron mass

7. Ratio of number of protons to number of electrons

8. Ratio of proton to electron charge

9. Expansion rate of the universe

11. Hoyle, "The Universe", 393.

12. Ross, The Creator and the Cosmos, 154–57, 188–93.

10. Mass density of the universe

11. Baryon (proton and neutron) density of the universe

12. Space energy or dark energy density of the universe

13. Ratio of space energy density to mass density

14. Entropy level of the universe

15. Velocity of light

16. Age of the universe

17. Uniformity of radiation

18. Homogeneity of the universe

19. Average distance between galaxies

20. Average distance between galaxy clusters

21. Average distance between stars

22. Average size and distribution of galaxy clusters

23. Numbers, sizes, and locations of cosmic voids

24. Electromagnetic fine structure constant

25. Gravitational fine-structure constant

26. Decay rate of protons

27. Ground state energy level for helium-4

28. Carbon-12 to oxygen-16 nuclear energy level ratio

29. Decay rate for beryllium-8

30. Ratio of neutron mass to proton mass

31. Initial excess of nucleons over antinucleons

32. Polarity of the water molecule

33. Epoch for hyper-nova eruptions

34. Number and type of hyper-nova eruptions

35. Epoch for supernova eruptions

36. Number and types of supernova eruptions

37. Epoch for white dwarf binaries

38. Density of white dwarf binaries

39. Ratio of exotic matter to ordinary matter

40. Number of effective dimensions in the early universe

41. Number of effective dimensions in the present universe

42. Mass values for the active neutrinos

43. Number of different species of active neutrinos

44. Number of active neutrinos in the universe

45. Mass value for the sterile neutrino

46. Number of sterile neutrinos in the universe

47. Decay rates of exotic mass particles

48. Magnitude of the temperature ripples in cosmic background radiation

49. Size of the relativistic dilation factor

50. Magnitude of the Heisenberg uncertainty

51. Quantity of gas deposited into the deep intergalactic medium by the first supernovae

52. Positive nature of cosmic pressures

53. Positive nature of cosmic energy densities

54. Density of quasars

55. Decay rate of cold dark matter particles

56. Relative abundances of different exotic mass particles

57. Degree to which exotic matter self-interacts

58. Epoch at which the first stars (metal-free pop III stars) begin to form

59. Epoch at which the first stars (metal-free pop III stars cease to form

60. Number density of metal-free pop III stars

61. Average mass of metal-free pop III stars

62. Epoch for the formation of the first galaxies

63. Epoch for the formation of the first quasars

64. Amount, rate, and epoch of decay of embedded defects

65. Ratio of warm exotic matter density to cold exotic matter density

66. Ratio of hot exotic matter density to cold exotic matter density

67. Level of quantization of the cosmic space-time fabric

68. Flatness of universe's geometry

69. Average rate of increase in galaxy sizes

70. Change in average rate of increase in galaxy sizes throughout cosmic history

71. Constancy of dark energy factors

72. Epoch for star formation peak

73. Location of exotic matter relative to ordinary matter

74. Strength of primordial cosmic magnetic field

75. Level of primordial magneto-hydrodynamic turbulence

76. Level of charge-parity violation

77. Number of galaxies in the observable universe

78. Polarization level of the cosmic background radiation

79. Date for completion of second reionization event of the universe

80. Date of subsidence of gamma-ray burst production

81. Relative density of intermediate mass stars in the early history of the universe

82. Water's temperature of maximum density

83. Water's heat of fusion

84. Water's heat of vaporization

85. Number density of clumpuscules (dense clouds of cold molecular hydrogen gas) in the universe

86. Average mass of clumpuscules in the universe

87. Location of clumpuscules in the universe

88. Dioxygen's kinetic oxidation rate of organic molecules

89. Level of paramagnetic behavior in dioxygen

90. Density of ultra-dwarf galaxies (or supermassive globular clusters) in the middle-aged universe

91. Degree of space-time warping and twisting by general relativistic factors

92. Percentage of the initial mass function of the universe made up of intermediate mass stars

93. Strength of the cosmic primordial magnetic field[12]

However fundamental constants are often coupled or related to each other so that fixing one constant's value fixes the other constants since they are related in a mathematical ratio or relation. So to avoid over-counting the number fine-tuned constants or parameters, scientists search for the fundamental constants that appear to be undetermined by any other constant or parameter. In this case, one can compile a list of over 20 fine-tuned constants. Reproduced below is such a list with explanations taken from a paper by Jay Richards[13].

COSMIC CONSTANTS

1. Gravitational force constant (large scale attractive force, holds people on planets, and holds planets, stars, and galaxies together)—too weak, and planets and stars cannot form; too strong, and stars burn up too quickly.

2. Electromagnetic force constant (small scale attractive and repulsive force, holds atoms electrons and atomic nuclei together)—If it were much stronger or weaker, we wouldn't have stable chemical bonds.

3. Strong nuclear force constant (small-scale attractive force, holds nuclei of atoms together, which otherwise repulse each other because of the electromagnetic force)—if it were weaker, the universe would have far fewer stable chemical elements, eliminating several that are essential to life.

4. Weak nuclear force constant (governs radioactive decay)—if it were much stronger or weaker, life-essential stars could not form. (These are the four "fundamental forces.")

5. Cosmological constant (which controls the expansion speed of the universe) refers to the balance of the attractive force of gravity with a hypothesized repulsive force of space observable only at very large size scales. It must be very close to zero, that is, these two forces must be nearly perfectly balanced. To get the right balance, the cosmological constant must be fine-tuned to something like 1 *part in* 10^{120}. If it were just slightly more positive, the universe would fly apart; slightly negative and the universe would collapse. As with the cosmological constant, the ratios of the other constants must be fine-tuned *relative to each other*. Since the logically-possible range of strengths of some forces is potentially infinite, to get a handle on the precision of fine-tuning, theorists often think in terms of the *range* of force strengths, with gravity the weakest, and the strong nuclear force the strongest.

13. Richards, "List of Fine Tuning Factors".

The strong nuclear force is 10^{40} times stronger than gravity, that is, ten thousand, billion, billion, billion, billion times the strength of gravity. Think of that range as represented by a ruler stretching across the entire observable universe, about 15 billion light years. If we increased the strength of gravity by just 1 *part in* 10^{34} of the range of force strengths, (the equivalent of moving less than one inch on the universe-long ruler), the universe couldn't have life sustaining planets.

INITIAL CONDITIONS AND "BRUTE FACTS"

6. Initial Conditions. Besides physical constants, there are initial or boundary conditions, which describe the conditions present at the beginning of the universe. Initial conditions are independent of the physical constants. One way of summarizing the initial conditions is to speak of the extremely low entropy (that is, a highly ordered) initial state of the universe. This refers to the initial distribution of mass energy. In *The Road to Reality*, physicist Roger Penrose estimates that the odds of the initial low entropy state of our universe occurring by chance alone are on the order of 1 *in* $10^{10^{(123)}}$. This ratio is vastly beyond our powers of comprehension. Since we know a life-bearing universe is intrinsically interesting, this ratio should be more than enough to raise the question: Why does such a universe exist? If someone is unmoved by this ratio, then they probably won't be persuaded by additional examples of fine-tuning. In addition to initial conditions, there are a number of other, well known features about the universe that are apparently just brute facts. And these too exhibit a high degree of fine-tuning. Among the fine-tuned (apparently) "brute facts" of nature are the following:

7. Ratio of masses for protons and electrons—if it were slightly different, building blocks for life such as DNA could not be formed.

8. Velocity of light—if it were larger, stars would be too luminous. If it were smaller, stars would not be luminous enough.

9. Mass excess of neutron over proton—if it were greater, there would be too few heavy elements for life. If it were smaller, stars would quickly collapse as neutron stars or black holes.

"LOCAL" PLANETARY CONDITIONS

But even in a universe fine-tuned at the cosmic level, local conditions can still vary dramatically. As it happens, even in this fine-tuned universe, the vast majority of locations in the universe are unsuited for life. In *The Privileged Planet*, Guillermo Gonzalez and Jay Richards identify 12 broad, widely recognized fine-tuning factors required to build a single, habitable planet. *All 12 factors can be found together in the Earth.* There are probably many more such factors. In fact, most of these factors could be split out to make sub-factors, since each of them contributes in multiple ways to a planet's habitability.

10. Steady plate tectonics with right kind of geological interior (which allows the carbon cycle and generates a protective magnetic field). If the Earth's crust were significantly thicker, plate tectonic recycling could not take place.

11. Right amount of water in crust (which provides the universal solvent for life).

12. Large moon with right planetary rotation period (which stabilizes a planet's tilt and contributes to tides). In the case of the Earth, the gravitational pull of its moon stabilizes the angle of its axis at a nearly constant 23.5 degrees. This ensures relatively temperate seasonal changes, and the only climate in the solar system mild enough to sustain complex living organisms.

13. Proper concentration of sulfur (which is necessary for important biological processes).

14. Right planetary mass (which allows a planet to retain the right type and right thickness of atmosphere). If the Earth were smaller, its magnetic field would be weaker, allowing the solar wind to strip away our atmosphere, slowly transforming our planet into a dead, barren world much like Mars.

15. Near inner edge of circumstellar habitable zone (which allows a planet to maintain the right amount of liquid water on the surface). If the Earth were just 5% closer to the Sun, it would be subject to the same fate as Venus, a runaway greenhouse effect, with temperatures rising to nearly 900 degrees Fahrenheit. Conversely, if the Earth were about 20% farther from the Sun, it would experience runaway glaciations of the kind that has left Mars sterile.

16. Low-eccentricity orbit outside spin-orbit and giant planet resonances (which allows a planet to maintain a safe orbit over a long period of time).

17. A few, large Jupiter-mass planetary neighbors in large circular orbits (which protects the habitable zone from too many comet bombardments). If the Earth were not protected by the gravitational pulls of Jupiter and Saturn, it would be far more susceptible to collisions with devastating comets that would cause mass extinctions. As it is, the larger planets in our solar system provide significant protection to the Earth from the most dangerous comets.

18. Outside spiral arm of galaxy (which allows a planet to stay safely away from supernovae).

19. Near co-rotation circle of galaxy, in circular orbit around galactic center (which enables a planet to avoid traversing dangerous parts of the galaxy).

20. Within the galactic habitable zone (which allows a planet to have access to heavy elements while being safely away from the dangerous galactic center).

21. During the cosmic habitable age (when heavy elements and active stars exist without too high a concentration of dangerous radiation events).[13]

"What in the world do you mean by coupled constants?" asked Mary.

Bert thought for a minute and then replied, "Well you can express some constants in terms of other constants."

"Huh??"

"Let me try this again," Bert quickly strode over to the dry erase board in the empty classroom in which they were talking. Grabbing a dry-erase marker in the board tray he wrote on the board the following equation and constants:

c = speed of light (electromagnetic radiation) in a vacuum

ε_0 = permittivity constant (or vacuum permittivity)

μ_0 = permittivity of free space constant

$$c = (\mu_0 \varepsilon_0)^{-1/2}$$

"Now notice that if I multiply the permittivity constant by the permittivity of free space and take the square root and then divide that into one, I get the speed of light. But the speed of light is one of the fine-tuned constants. Can I argue that ε_0 the permittivity constant is fine-tuned if the evidence leans that way? Can I argue that μ_0 the permittivity of free space constant is also fine-tuned if the evidence leans that way? Not necessarily because if c the speed of light is fixed then the values of μ_0 and ε_0 are somewhat determined also (or at least the ratio of μ_0 to ε_0). So calling each one a separate fine-tuned constant is not necessarily fair. But if you could show that the ratio between are μ_0 and ε_0 are critical, then you could get another fine-tuning constant or parameter out of that, but no further." Bert turned to face Mary with a questioning do-you-get-it look on his face.

"What the heck are the vacuum permittivity constant, and the permittivity of free space, and are they fine-tuned also?" Mary questioned.

"I better back up. No they are not known to be fine-tuned at this point. I was just using this as an example equation to illustrate the idea of coupled constants. The equation couples these constants in some way, so that any fine-tuning in one carries over in some way to the other constants if it is there. Just substitute a, b, and c in for these constants if their names bother you, it's the idea that counts."

Mary pondered the board a bit, "Okay, now I see it. "That is why Hugh Ross can come up with a list of 93 fine-tuned constants, and Jay Richards only comes up with 21 fine-tuned constants. Well is Ross all wrong in his list then?"

"Not necessarily," explained Bert, "there may be some fine-tuned ratios in his longer list. He is just taking a more liberal approach to his fine-tuning list than Jay Richards. If you wanted to compute an overall probability by just multiplying the probabilities of each fine-tuned constant together, Jay Richards' list would probably be more accurate."

"But if you want to impress somebody with the extent of fine-tuning design in the physical world, Hugh Ross' list is more impressive," interjected Mary, "But where did you learn so much about physics? I thought you were a philosophy major?"

Bert never one to stop a conversation with Mary, replied, "Well I was a physics major before I switched to philosophy, who knows, I may double major, if I can stand the extra stress."

Mary added, "Did Maxwell actually predict the speed of light before they could measure the speed of light?"

"No they had some pretty good measurements of the speed of light by 1865 when professor James C. Maxwell published his classic work predicting radio waves, that radio waves were made up of electrical and magnetic fields, and that light was just one example of this type of "electromagnetic wave." But when his equation showed that the speed of his electromagnetic waves were the same as the measured speed of light, he made the obvious next step of suggesting that light *was* a type of electromagnetic radiation. He was right and has gone down in history as the third most influential physicist after Newton and Einstein. Maxwell was trying to unify electricity and magnetism and succeeded magnificently. What is very interesting is that his entire work was based on the concept that light, electromagnetic waves, radio waves like all other waves had to oscillate and travel in some medium, which scientists of the day called the luminiferous ether.

"What the heck is that?!" cried Mary.

"Well sound waves have to travel in air, (in outer space there is no sound). Water waves have to travel in water, so light waves or electromagnetic radiation has to travel in some kind of medium that we cannot feel or sense," explained Bert.

"So?"

Bert paused and then said, "Well we now know there is no such thing as the luminiferous ether. Light, radio waves, electromagnetic radiation— do not need any medium to propagate in, it just propagates. But Maxwell used it, to set up his theory of electromagnetism."

"Haven't they had to adjust or fix his theory to make up for the fact that his basic assumption on the nature of light and how it travels was fundamentally wrong?" asked Mary.

"No, that is the crazy thing about it all. They have condensed his equations from several down to four basic equations, and modernized the notation. But the equations work, with or without the luminiferous ether. He (and everybody of that day) had a very incorrect notion of how light worked, but created a mathematical theory that works like a charm to this day. It's not like his conception of light was all wrong, it just had a very basic flaw that did not become obvious until forty years later." Bert grinned after that statement.

"Well what other basic theories of science have their foundations based on faulty fundamental assumptions?" queried Mary.

Bert grinned even bigger, "Not a good question to ask if you are seeking fame and fortune in the scientific public square. That is a backroom coffee break question that you forget after you return to work. But philosophically it is a very valid one. It turns out that many if not most scientific disciplines had their start in some very flawed beginnings."

"Like what?"

"Thermodynamics—the field started out treating heat as a fluid that flowed from hot to cold objects, not the motion of atoms and molecules that we know it to be today."

"And?"

"Newtonian mechanics, Isaac Newton proposed this bizarre notion that a mysterious force (gravity) acts instantly all over the universe drawing all masses together. He was so bold as to not even bother to try to explain it. He just said, "I feign no hypothesis." "

"And it is wrong"

"Yes" said Bert, "Einstein has replaced it with a theory that gravity is just masses sliding down the curvature of space toward the local bottoms of the space-time curvature of space in four dimensional space-time."

"And this is better?" gasped Mary.

"Oh, absolutely, mathematically we can predict the motions of objects, planets, galaxies; better, explain time dilation, and so on. But you do have a point. Have we really removed the mysterious element, the fundamental assumption that could be flawed to the core from it all? I really don't know, no one does, but God. What really is space-time curvature? How do I know that it really exists? Am I building a really successful theory on a bad a priori assumption?" Bert paused and grew quite introspective. "I guess you buy a particular scientific worldview because the theory you get your worldview from is so successful in doing things for you. But that is still no guarantee that there isn't something fundamentally wrong with the worldview you derived from your successful theory."

"But couldn't the same be said for other academic disciplines? History, theology, language studies?" Mary eyed Bert carefully.

"Yes, it could. All of the academic disciplines are caught in similar dilemmas. I guess you could say science isn't any better or worse, just caught in the same sort of issues and dilemmas; just part of the grand struggle of humanity to understand the questions of their existence," Bert replied.

"Then why is science placed on such a high pedestal in our western culture?"

Bert smiled, "Because historical studies did not give us iPhones."

We would like to think that crass materialism, love of gadgets, comfort, and convenience, is not the deciding factor in shaping modern western philosophy and thought. But it would be foolish to think that it has no role. But there is the counter argument; iPhones and convenience do not take away the pain of a failed marriage, the death of a child or loved one, and dealing with the meaninglessness and drudgery of many modern "careers." Most people look for meaning in their work and lives through their families, and their religious beliefs. At his point the incredible fine-tuned constants of the universe act as pointers; concrete scientific pointers that *something* beyond humanity designed the universe so that we could live and exist. It also drops hints as to the nature of that something. Fred Heeren in his book

Show Me God[14] argues that they point to a God who is independent, infinite, and personal. His chain of logic goes like this:

The universe is an effect which demands a very great cause
A series of causes cannot be infinite. There must have been a first cause, which itself is uncaused. (But what caused God? Silly question, He is uncaused. The same question could be asked for an eternal universe. What caused it?)

The first cause must be independent of its effect
The first cause must not require any of the things that depend on *it* for *their* existence. This does not bode well for Star Wars-like impersonal forces that control the universe, since that force depends on life or something in the universe for its existence. The first cause must be *above* the universe or transcendent to it. If the universe shows signs of incredible design or *mind*, then that implies a mind that transcends the universe.

The first cause must be infinitely powerful (omnipotent)
The first cause had to be unlimited, because if it were limited, it would have to be limited by some other thing and it wouldn't be completely independent any more. This introduces the logically possibility of miracles, as infinite power can do as it wishes, miracles or not.

The first cause must be spiritual (transcending space)
We must deduce that an entity outside the universe is the only kind that could have created it.

The first cause must be all-knowing (omniscient)
It is reasonable to assume that the creator/first cause of all that is *knows* all about what it created.

The first cause must have personhood
It is reasonable to infer that the first cause/creator who displays purpose and intention in creating a universe fine-tuned for life. The most important attribute of personhood is purpose, thus grand purpose implies grand personhood.

Heeren is drawing from the classical Kalam cosmological argument in making a case for a personal God from the cosmological data. The Christian

14. Heeren, Show Me God, 88–96.

philosopher and apologist William Lane Craig[15] has discussed this argument in detail. Hereen also answers a question posed by the astronomer
Fred Hoyle who cited a modern theologian in asking, "What we cannot
understand is that God who has no need of the world should have reason
to create (it)." This is a very common reaction to God from those who are
skeptical of His existence. It betrays a common trait among skeptics, the
tendency or need to make the first cause/God less than what God really is.
It is a type of straw man argument. You create a familiar image of God, but
one that is crippled in a subtle but important fashion. Hereen answers the
question; God is not a *Thing*, but a *Person* who *desires* and is *pleased* to create
a universe with beings such as ourselves who can discover and commune
with Him.

Materialists respond to the argument for design from fine-tuning in
several ways. One way is to argue that we just don't know enough yet, given
enough time we will eventually figure out how it all works without any kind
of divine intervention. We hope to figure it all out materially someday, because, "see how well we have removed past accepted cases of divine intervention in earth's history and actions." But his type of thinking is similar
to theists who sometimes argue that the unknowable, the obvious gaps in
our knowledge can sometimes be attributed to God's inscrutable action or
miraculous hand. This is often derided as a *God of the Gaps*, argument. God
fills in for all our gaps in our knowledge. *God of the Gaps* is often called a
science stopper. Why bother to investigate something for material causes
if we have already deemed that God caused it and therefore no scientific
explanation can be found. Now apply the same sort of thinking in reverse
to the, "we will eventually figure it all out argument," or *Materialism of the
Gaps* argument. What if it really was designed by some transcendent agent?
Then continually looking for a material cause will distract from the real
business of figuring out the tell-tales signs of a transcendent agent, who
guides creation in some fashion, how is it done, and when can we declare
the agent's guiding hand, and when can we declare that material causes are
sufficient to explain it? At this point it becomes very obvious that one's a
priori assumptions about what is science and what is the nature of reality will make a huge difference in how one looks at the data. You can also
argue that the relationship between the fine-tuning data and one's a priori
assumptions can often be coupled; as Fred Hoyle's comment cited earlier,
"*a common sense interpretation of the facts suggest that a super intellect has
monkeyed with physics, as well as with chemistry and biology, and that there
are no blind forces worth speaking about in nature,*" strongly implies. The

15. Craig, The Kalaam Cosmological Argument.

force of the fine-tuning data can change one's a priori assumptions. Hoyle though he remained an atheist/pantheist to the end, would never have made such a comment before the discovery of fine-tuning in the universe. Others scholars such as Anthony Flew converted from atheism to theism[16] after being exposed to the data.

Another response to the fine-tuning argument is to suggest a multiverse. A multiverse is an ensemble of universes that are continually birthed, and formed. They are caused by quantum fluctuations, observed events, quantum branching, or whatever. Every possible universe that can be imagined is there as some sort of parallel universe. In one universe you married Elvis Presley. In another you never finished high school and died a horrible death at age fifteen. The vast, vast majority of these universes are very dull and boring rearrangements of trillions and trillions of hydrogen molecules. But since you have forever to go through every possible rearrangement of the universe, you will ever so rarely; oh, ever so rarely get one with the fundamental constants just so, so that you get life and you. Voila! The problem of the fine-tuned constants is explained; we lucked ourselves into it, otherwise we wouldn't be here to observe ourselves. We did have forever, didn't we? The idea is we would not be here to find these fine-tuned constants if we did not have them. This idea is commonly called the weak anthropic principle. But there is a major problem with the multiverse hypothesis, probably many problems, but we will discuss just one. In all of the universes created the majority of them will not have significant rearrangements of matter in things as complex and ordered as livings plants, animals, and human beings. If you push a pile of clay dirt down a steep embankment, by the time the dirt pile reaches the bottom, you usually have a random looking pile of clay dirt at the bottom, not an upright clay dirt replica of the Statue of Liberty staring at you. (Chemists and physicists have a name for this idea, it is called entropy.) But if you had forever you could argue that the Statue of Liberty in clay might show up at least once. But you would have an awful large number of random piles of dirt for every Statue of Liberty. But what are the odds of getting a square pile of dirt instead of the Statue of Liberty? It would be pretty freaking rare, but definitely more common than a Statue of Liberty. Okay then; in these forever sets of universes which is going to be more common—a human being formed with a family, a history, from an incredible set of fine-tuned coincidences, or a brain that randomly comes into existence in the sea of hydrogen molecules, complete with a set of memories of growing up with a family, and a mental state that it is staring at a book and reading it. This is sometimes called a Boltzmann brain after Ludwig

16. Flew, There is a God.

Boltzmann who in the late 1800s suggested such a possibility. Oops! That brain just died in the empty vacuum of space. But just wait a forever until another universe comes along to spawn another Boltzmann brain that is just one page further in your reading, complete with memory of the last page of reading. But we'll have to skip the universe where you are Elvis. Around 2008 cosmologists realized that Boltzmann brains were far more likely than actual fine-tuned universes like the one in which we live. Thus they would have to argue that: you reading this page is not a real you in a real fine-tuned universe, but a Boltzmann brain . . . At this point one wonders what is left of rational thought, logic or anything if you are going to keep thinking this way. Perhaps that is why the NY Times reporter writing on this discovery entitled his article, *Big Brain Theory: Have Cosmologists Lost Theirs?*[17] This idea is a major blow to the evolution of multiverses, and even the very idea of multiverses. If an individual continues to pursue lines of thought along these lines, you can legitimately question whether their a priori beliefs about the nature of the universe are more important than their rationality.

Now let's finish up on the anthropic principle. There are many different formulations of it, but we can break it down into four basic categories.

Weak Anthropic Principle (WAP)—We see that the universe was fine-tuned for our benefit because if it were not, we would not be here.

Strong Anthropic Principle (SAP)—The ending result of the universe was to produce us, so the various fine-tuned physical constants and other properties which are critical to our existence have to be such that they bring about our existence. The iconoclastic physicist Freeman Dyson expressed it as, "*As we look out into the universe and identify the many accidents of physics and astronomy that have worked together to our benefit, it almost seems as if the universe must in some sense have known we were coming.*"[18]

Participatory Anthropic Principle (PAP)—Because quantum states cannot be resolved into a reality without an observer, then the universe could not exist without observers. Reality is represented as a wavefunction in quantum mechanics. The traditional Copenhagen interpretation of quantum mechanics argues that a wavefunction is indeterminate until an act of observation is made on it, thus no observers, no universe.

17. Overbye, "Big Brain Theory".

18. Dyson, Disturbing the Universe, 250.

Final Anthropic Principle (FAP)—intelligent beings and life must come into existence in the universe, and once it does it will never die out. (Perhaps we evolve and become God?) This final anthropic principle is sometimes ridiculed as the Completely Ridiculous Anthropic Principle.

The WAP is often used by agnostics and materialists to minimize the theistic import of fine-tuning. The SAP is used by theists to support the notion of God, and by Pantheists, Deists, Christians, Jews, Muslims, or just interested others. The PAP tends to be supported by Pantheists, or those with an over-grown sense of importance; and the FAP and it variants, Gaia, etc., are embraced by New-Agers and all sorts of people. It is wise to find who your buddies or co-embracers of a particular anthropic principle are, before using it to stake out a position. But if you want to shake the faith of an uninformed atheist, the theistic ramifications of fine-tuning in the universe are a good place to start.

In the pivotal text *The Privileged Planet*[19], astronomer Guillermo Gonzalez and science philosopher Jay Richards postulated the privileged planet hypothesis. We can summarize the idea as follows, those places in the universe that are most suited for life are also the places in the universe that are best for scientific discovery or finding things out about the universe. Or to phrase it more pithily, *the most habitable places in the universe are also the most measurable ones.* It is as if "Someone," who has placed us in the most habitable place in the universe also wanted us to be in the best possible position to discover what had been done and discover best the universe around us.

To the theist, that "Someone" is God, and it makes perfect sense that in finding the safest and best place for humans to live, He would also put us in the best place to discover his handiwork. It would be difficult to find a better rationale for pursuing scientific research on our universe. If God has put us in one of the best places to survey and understand our solar system and universe, then we had better get busy and see what He has in store for us. However to a materialist or deist who does not see God's personal interaction in the world, the idea is complete nonsense and utterly egotistical. This may explain the reaction Gonzalez and Richards got, when an hour long movie justifying their idea was first shown. They rented Baird's Auditorium in the Museum of Natural History at the Smithsonian Institute in Washington D.C. to premiere *The Privileged Planet* movie[20]. The museum staff reviewed the movie and agreed to let the Discovery Institute

19. Gonzalez and Richards, The Privileged Planet.
20. Privileged Planet.

use their logo on the advertising for an invitation only group of about 200 people. After the invitations went out, the Smithsonian was bombarded by calls from disgruntled Darwinists, materialists, scientists, who wanted the Smithsonian to withdraw their co-sponsorship. Under the intense pressure, they withdrew their co-sponsorship but still allowed the film to be shown as scheduled. One viewer, PhD biologist, Ray Bohlin[21] stated that the movie was a great scientific film, and he was completely mystified as to the fuss. He then commented, "Just remember science is pursued by *people*, and everyone has a worldview that can alter dramatically how science is perceived and what counts as science."

How did Gonzalez and Richards support their privileged planet hypothesis? We can briefly summarize some of their ideas. If the best places to live in the universe are the best places to discover the universe then what "habitability criteria" gives the best discoverability or measurability criteria? Here is a brief list gleamed from Gonzalez and Richards[22].

Habitability for Life	*Resulting Discoverability and Measurability*
Moon	Information given by solar eclipses
Circumstellar Habitable Zone	Discoverability of the solar system
Galactic Habitable Zone	Discoverability of our galaxy and beyond
Atmosphere of N2, O2, CO2	Transparent to visible light for discovery
Geologic data recorders	Chemical cycles necessary for life
Plate tectonics	Discoverability of earth and its past
Carbon & oxygen needed for life	Understanding earth's past
Cosmic time and habitability	Discoverability of the universe

Gonzalez and Richards start *The Privileged Planet* book with remarkable coincidences involved with total solar eclipses on earth by the moon. The apparent diameter of the moon by an observer on the earth, just matches the apparent diameter of the sun by the same earth-bound observer. This is what makes total solar eclipses so spectacularly beautiful *and* scientifically interesting. The very bright chromosphere of the sun is just covered up, but the very, very thin corona atmosphere around the sun, is not covered up. That 1– 3 million degree hot pearly pink circle of light gives us the beauty of a solar total eclipse. The sun is 400 times larger than the moon, but being 400 times further away, the apparent diameter just matches. The match is so fine-tuned that on rare occasions when the moon orbiting the earths is near

21. Bohlin, "An Unwanted Premiere!"
22. Gonzalez and Richards, The Privileged Planet, 18.

its furthest point, (apogee), from the earth it doesn't completely cover the sun during a solar eclipse, and all you see is a ring of sun (annular eclipse), but no pink corona. Because the moon is moving away from the earth at 3.82 centimeters[23] per year, this fortuitous match has been going on for some past millions of years and for only the next 250 million years. This is just 5 percent of the earth's existence if you accept the old earth hypothesis. And this during the same time period that humans have been on the earth to study it. This is odd.

Well what did solar eclipses contribute to science? The first photographs of the July 28, 1851 total solar eclipse allowed George Airy to describe the sun's chromosphere. Previous solar eclipses had yielded spectra of the chromosphere that revealed a new element, helium, which was not verified in the laboratory until 1895. During the December 22, 1870 total solar eclipse, American astronomer Charles Young discovered the flash spectrum of the sun. This revealed how atomic spectra are formed in the sun and thus by parallel, how they are formed in distant stars. Spectroscopy of the stars has been our key to unlocking the mysteries of the cosmos. In November of 1915, Einstein published his theory of general relativity that predicted light rays would deflect under the influence of intense gravity. Astronomers Arthur Eddington and Edwin Cottingham traveled to the site of the May 29, 1919 total solar eclipse to photograph the positions of stars near the sun during the eclipse. The resulting stellar deflections confirmed the predictions of Einstein, and made him an international hero. Much of modern astronomy got its start from critical scientific observations made during solar eclipses.

Circumstellar Habitable Zone (CHZ) is the distances from the sun or any other star that produces temperatures on a planet that allows water to stay a liquid. If we move the earth into a closer orbit like Venus, the resulting heat from the sun would vaporize all of the earth's water. If we move it into a more distant orbit like Mars, all of the water on earth would freeze. Some estimates put the earth's habitable zone at 95 percent of our present distance from the sun to 137 percent of our distance from the sun. Less than 95 percent of the earth's distance from the sun, the earth would experience runaway evaporation of water on the planet, and at more than 137 percent, runaway freezing. Neither of these alternatives would be conducive to the development of life on earth. With a less bright sun, we would have to move the CHZ and the earth closer in to support life, but that brings a host of other problems into play (tidal locking of the earth's rotation, sun's intense radiation blasting away our atmosphere, etc.). By making the sun more

23. Ross, Improbable Planet, 212–19.

bright and warmer we would have to move the CHZ and the earth further out, but that creates problems also, (more frequent life killing asteroid impacts from the asteroid belt, a much longer year causing massive changes in our seasons, gravitational perturbations from Jupiter that could upset the singular stability of earth's orbit). However our present earth-moon-sun system has provided very interesting ways to estimate the distances between the three bodies that goes back to the time of the ancient Greeks. Eratosthenes of Cyrene calculated the earth's size by using the Sun's zenith positon at two widely separated points on the earth's surface around 200 BC. In 189 BC Hipparchus of Nicaea used the degree of coverage of the sun by the moon during a total solar eclipse at two different cities to estimate the distance to the moon. In the third century BC, Aristarchus of Samos devised a geometrical method using angles between the half phase moon and the sun to estimate the distance to the sun. Even though he was way off, due to the error in his angle measurements, it allowed him to estimate the sun was 20 times as far away as the moon (actual value was 390 times further away), and thus reason the sun had to be much larger than the earth and the moon, and thus ought to be at the center of the solar system instead of the earth. Thousands of years later his writings influenced Nicolas Copernicus to propose his sun centered solar system and launch modern astronomy. The earth's distance from the sun allows the development of a thin sheath of nitrogen and oxygen to envelope our planet, thus allowing life, and allowing that life to peer out through the transparent air to see the starry heavens in the most information rich section of the electromagnetic spectrum, the visible light region.

Our sun and solar system are perched about 2/3 of the way out from the center of our Milky Way galaxy, in between two spiral arms of our galaxy. Any closer into the core of the galaxy and the higher rate of super novae, and gravitational perturbations of closer nearby stars would upset the planetary orbits about our sun. So either by radiation or erratic planetary orbits; the sun's position in the galaxy would sterilize life on earth. The view closer to the core of the galaxy would be so crowded with nearby stars and galactic dust that we would likely never discover other galaxies. If our sun was further away, the rate of stellar novae would have been insufficient to produce the heavy metals spewed through space that are necessary to form rocky planets with metallic cores, so necessary to sustain life on earth. No hot metallic core with convection currents, and the earth would have no plate tectonics, no continents, no mountains, no carbon cycles, no sulfur cycles, no volcanos, and no hydrologic cycles to sustain advanced life. But further out would increase our view of other galaxies, but greatly decrease our view of our own, particularly the Milky Way core. If we were located in a spiral arm we may have never realized our Milky Way galaxy had spiral arms, and we would have been a more crowded stellar neighborhood with a higher chance of nearby life killing super novae. But being positioned between two galactic spiral arms we avoid these dangers, while getting a grand view of both arms, and the core of the galaxy in one plane. Then by looking perpendicular we have a clear view of the rest of the universe with its myriad of galaxies and super clusters. Once again habitability and discoverability appear to be in the same place.

Carbon, oxygen, nitrogen, and hydrogen chemistry appears to be the only chemistry capable of giving us the necessary complex suite of organic compounds needed for life. Science fiction aside, silicon or any other type of chemistry just doesn't give us the needed complexity to store information and make molecular machines. Most scientists have realized that you need carbon, oxygen, nitrogen, hydrogen, and a few other elements; if you want to have life. Carbon and oxygen have elemental isotopes with ratios between them that have proved invaluable in approximating temperatures, and the presence of past life, and dating carbon based objects of less than 20,000 years. We can unearth an old fire pit, or wigwam, and tell you if it was built 100 years ago, or 10,000 years ago by checking the carbon 14 isotope content. Every summer or winter another layer of snow is deposited on Antarctica and Greenland, compressing the previous year's annual layer. By drilling and extracting an ice core of the snow pack in Greenland or Antarctica, one can freeze it and count the layers back to ~250 million years ago in Greenland, or ~900 million years ago in Antarctica. By extracting an ancient air bubble trapped in the ice, you can date it by its position in the ice core column and then measure the oxygen isotope ratios to gain an approximate measure of the earth's global temperatures. This has yielded enormous data on the earth's average temperature over the ages, the glacial cycles, and other rich ancient information. It has also brought to light our very our fortuitous position in time between two global ice ages that has allowed our current technological civilization to flourish.

So you could argue that our position in the solar system, our position in the galaxy, that gives us the best measurability (discoverability) for the universe also provides us with the right elements on the surface of the earth, and right elements in the core of the earth, provide us with the seasons, magnetic field, plate tectonics, atmosphere, geologic cycles, and chemical cycles that are crucial for our existence (habitability). The best places to live are the best places to discover. Why is this so? Just an accident?

If one accepts an old earth perspective the issue becomes even more complicated. We are positioned in time for the best possible view of the universe. Too early and the first generation of stars in the universe have not had enough super novae to generate enough heavy metals to sustain life, and the galaxies have not reached the right stage of galactic development to support life giving star systems; but 14.5 billion years is about right. If we wait in cosmic time to many billions of years later; the universe will have expanded to the point that we cannot see the far reaches of the universe with our telescopes. On a more recent time scale, it appears that the last nine thousand years on the earth have been the *only* nine thousand years where our climate[24] and sea levels have been at the most stable point that an advanced global civilization could travel to and fro on the seas, and flourish. If we push the privileged planet hypothesis this far then you could argue that: The best places *and times* to live are the best places *and times* to discover.

But before we leave this idea a warning rejoinder from the French essayist M. Montaigne[25] (1533–92) is needed.

> Why should not a gosling say thus: all the parts of the Universe regard me: the earth serves me for walking, the sun to give me light, the stars to inspire one with their influences. I have this use of the winds, that of the waters; there is nothing which this vault so favourably regards as me; I am the darling of nature. Does not man look after, lodge, and serve me? It is for me he sows and grinds: if he eat me, so does he his fellow-man as well: and so do I the worms that kill and eat him . . .

It is easy to get carried away with the design argument and how it is all so arranged to support humanity. Careful analysis and judgment of the data is necessary to separate the wheat from the chaff; and the texts from Gonzales and Richards[26], Hugh Ross[27],[28], and Barrow and Tipler[29], give the necessary supporting arguments for their fine-tuned arguments and privileged planet hypothesis.

24. Ross, Improbable Planet, 212–19.

25. Montaigne, Essays, 51.

26. Gonzalez, The Privileged Planet.

27. Ross, The Creator and the Cosmos.

28. Ross, Improbable Planet.

29. Barrow and Tipler, The Anthropic Cosmological Principle.

5

Cells, Machines and Biochemistry

DR. DELZER WAS AN interesting professor, a very good electrical engineer with slightly long hair and a physique for which women would die. He was nicknamed Dr. Death for his exams, and the end result of those who tried to keep up with him on his marathon runs. The black wavy hair and tiny goatee rounded out the effect. Approaching the age of fifty caused him to have a slight sense of desperation regarding his youth and vigor; that displayed itself in interesting ways in the classroom. Perhaps that was why his apologetic theology/science lectures had an enduring blend of heavy metal rock, high mountain vistas, and the role of science and engineering (mainly engineering), when he gave them.

"God is an engineer, an exquisite accommodating engineer. You see signs of it everywhere as you examine His creation. But how do you find that out? And what does this mean as you approach the Genesis record in the Bible," at that point Delzer queued the music, and heavy metal blasted the front row of the lecture room as the high Himalayan vistas filled the screen from the computer projector mounted on the ceiling. Seeing the wild-eyed look of the front row of girls, he quickly turned the volume down as the PowerPoint show moved to pictures of far galaxies flung across the sky, and the Hubble telescope deep field shots.

"Now let's look at the Genesis account and compare it with what we know from science about the earth's origins," pausing a bit, Dr. Delzer picked up his Bible and started reading. "I am reading from the Revised Standard Version, "*In the beginning God created the heavens and the earth. The earth was without form and void, and darkness was upon the face of the deep and the spirit of God was moving over the face of the waters. And God said, "Let there be light"; and there was light. And God saw the light was good; and God separated the light from the darkness. God called the light Day, and the darkness he called Night. And there was evening and there was morning, one day." "*

Looking up at the class he saw familiar faces, Mary, Ellie, Bert, Jon, Charles, Kristina, and Jack. "The creation story in the Bible is the only ancient creation story that posits creation *ex nihilo*, that is creation out of nothing, and the Big Bang theory of cosmology says exactly the same thing. After the heavens and the earth are created, there is a frame shift from God's perspective to an earthly (man's) perspective, both in space and time, to where the earth is dark and covered with water. This fits well with major theories on the earth's early history where the atmosphere was thick and dense, and water possibly covered the entire planet. Then as the early earth was transformed by the changing sun and solar system, the thick atmosphere was gradually blown away, allowing light to penetrate to the earth's surface, and concept of Day and Night became possible on the earth."

Ellie looked up and pondered, "So far he is following the line Dr. Hanson told us last semester."

"Now note the ending again, *And there was evening and there was morning, one day.* The Hebrew word *yom* is used with a number, an ordinal. Many young earth creationists argue that it has to be translated as a literal 24 hour day, as the Hebrew word for day, *yom* with a number attached, is only used in that context. But many Hebrew scholars disagree. They agree that Hebrew vocabulary is very limited and has no word for long periods of time, thus *yom* is often used to denote long periods of time, since there is no other way. Furthermore, *yom* with an ordinal is used elsewhere in the Bible to mean something other than a literal 24 hour day."

Ellie and Kristina looked up with a start, "Where in the Bible is that?" blurted out Ellie.

"Hosea 6:2 and Zechariah 14:7; and also note the ending *there was evening and there was morning* brackets a night, not a 24 hour period, suggestive of ending a long period of time. Now let's look at the next day," Dr. Delzer continued.

"*And God said, "Let there be a firmament in the midst of the waters, and let it separate the waters from the waters." And God made the firmament and separated the waters which were under the firmament from the waters which were above the firmament. And it was so. And God called the firmament Heaven. And there was evening and there was morning, a second day.*" Gesturing slightly with his right hand in an arc, Delzer explained, "The early Babylonian cosmology had the universe as follows, the earth as a flat round pancake like surface with waters below it (the waters of the deep) and the sky as a big bowl like structure held up over the earth. The waters above the firmament were above this upside down bowl and occasionally the heavens opened up a bit and let some of it fall through as rain. These verses seem to reflect that early cosmology, but they also reflect a very modern cosmology also. If we allow the *evening and there was morning* to denote a very long period of time, which the Hebrew easily allows, it fits nicely with modern theories. These theories have the earth's atmosphere clearing, the formation of the oceans, and the modern hydrologic cycle starting when the earth's plate tectonics kicks in and creates the first continents. It also fits well with verses 9—10 in day three where God causes the dry land to appear. But there is a problem here that we need to deal with. Any ideas class?"

Dr. Delzer paused and looked up and around at the class, "Ellie."

"But how can you be sure it refers to a long period of time? Historically the church over the last 1800 years has always viewed the *yom* in Genesis one as a literal 24 hour day. It's only in recent times we have chosen to alter that and just because of scientific discoveries," Ellie replied.

"Ellie, you are largely right, but not completely, some ancients Bible scholars thought seven days was too long and God created instantly as His omnipotence allows, and a few (very few) tied it to longer periods of time; but for the majority you are correct. But our new discoveries in understanding Hebrew in recent years have mirrored some of our rapid advances in scientifically understanding our past, so it is not surprising that these newer readings have come only recently. We have not needed them until now."

Jack piped up, "But with a little fudging you could get a lot of our recent geologic theories over the past 150 hundred years to match up nicely"

Delzer with a grin replied, "True, but not as nicely as you would suppose. The general modern trend of geologic evolution of the earth and appearance of sea, land plants and animals has stayed fairly consistent over the last 150 years. But you are still not seeing my problem. What if you read a newly discovered Shakespearean play and in the play they described a Ford Model T automobile?"

Bert answered this time, "You have a forgery, a fake. Ford Motors did not exist in Elizabethan times, and automobiles certainly did not."

"Correct, you have a historical anachronism on your hands, a sure sign of a bad fake, written in fairly modern times, a modern invention or concept written incorrectly into the past. Are we imposing a modern perspective, a scientific one onto an ancient manuscript, a historical anachronism that we have plastered over an ancient Hebrew worldview that doesn't even know what the scientific method is?" Dr. Delzer was grinning very broadly now.

The class was silent, carefully digesting this. Bert raised his hand and was about to speak, but Dr. Delzer who did not see him continued on.

"Maybe yes, but maybe no. I would say no."

"Why?" queried Bert.

"Look at Isaiah 53, verses 1 through 9, and Psalm 22 verses 1 through 18. Here David and Isaiah are predicting the manner of the Messiah's death in a detail that is reflected in the New Testament to an extraordinary degree. Death by crucifixion was not even known in the time of David and Isaiah, but here they are writing about the future Messiah's suffering in a manner that is abundantly clear to us today. Is that an historical anachronism?"

Bert stammered, "But that is prophesy!"

"And Genesis one is not prophesy in any form or fashion?"

"I don't know," Bert replied.

Delzer smiled, "And neither do I or anyone else for that matter. So it could reflect modern science as well as an ancient Mesopotamian worldview also. How accommodating and wise of God to speak to two (or more) completely separate cultural worldviews at the same time with a single origin story written long ago. To the extent we "get" or find the truth of our origins, the better will be the match between Genesis one and our modern theories. Thus I argue it is not a historical anachronism or imposing a modern worldview on an ancient text. Truth is truth no matter how ancient it is or how it is couched in the language of another culture."

Mary and Jon looked stunned. The rest of the class was stirring and chatting among them.

Continuing Dr. Delzer explained, "God is infinite and we are not, thus God has to accommodate to us to even speak to us. Why bother to speak when we could just think God's thoughts after Him and not even use language. But as the great reformer John Calvin wisely saw, God has to accommodate his way of speaking to our finite minds. He had to accommodate to the ancient Hebrews, the Romans and Hebrews of Jesus's day and to us even today. Do you really think we can fully understand all the details of how God created the world with our puny modern science?"

"How many different time periods or cultures has God accommodated to when Genesis was written," asked Bert.

"Ah! Good point, I really don't know. We could really get carried away with this, but I do think it is silly to *require* God to speak to only one culture through Genesis one, as some Hebrew linguists would argue. Now moving on to verse 11 through 13, "*Let the earth put forth vegetation, plants yielding seed according to their own kinds, and trees bearing fruit in which is their seed, each according to its kind. And God saw that it was good. And there was evening and there was morning, a third day.*" This also fits in with the development of plant life in modern theories and the fossil record. Now bear in mind we could argue a God directed evolution—but most Christians with the Day-Age or Old Earth Creationism perspective, have major problems with evolution. Thus they see God creating in stages or progressively over the eons, since a thousand years are but a day in God's sight. This is often called Progressive Creation and matches the Bibles nicely with the fossil record. The fossil record is not very gradual in its change, but marked by long periods of stasis or no change followed by the disappearance of species, genera and families, and the abrupt appearance of new ones. One of the dirty secrets of the fossil record is how few transition fossils have been really found. There are a few, and a couple of sequences, but they are surprisingly rare. Nothing at all like what Charles Darwin originally predicted.

Kristina interrupted, "But isn't that why Professors Niels Eldredge and Steven Jay Gould of Harvard postulated the punctuated equilibrium theory of macroevolution, to account for the scarcity of transition fossils?"

"Ah yes, I am going to get to that eventually. But they did that in 1972, and in my opinion injected some severe conundrums into the whole theory of macroevolution. But I think they had no choice. The fossil record by 1972 was causing some paleontologists to demand that something be done to fix Darwin's gradualism theory of macroevolution."

"Now moving on to the rest of the creation days we have in day four the appearance of the stars, sun and moon," Delzer was pointing to the verse in his Bible with his index finger. "*And God said, "Let there be lights in the firmament of the heavens to separate the day from the night: and let them be for signs and for seasons and for days and years, and let them be lights in the firmament of the heavens to give light upon the earth" And it was so. And God made the two great lights, . . . And there was evening and there was morning, a fourth day.*" Dr. Delzer paused, looked at the class and continued, "It is here that the average modern scientist parts company with our account. The idea of the sun, moon and stars being created after all of the rest of creation up to this point seems ridiculous. But the conservative Old Earth Creationist theologian or scientist would argue that the sky, and cloud cover which allowed for day, and night, has now cleared so that the sun, moon and stars could be visible from the surface of the earth for the first time. It does not square with the Mesopotamian Cosmology reading; but it will if you don't argue God has to be speaking *only* to ancient Hebrews in this text."

"Now in day five God creates the birds and the fishes, which places the birds way too early for the fossil record unless you count the winged dinosaurs, which the Hebrew is flexible enough to allow. Then on the sixth day God creates the animals and then man. Included with the creation of man is the creation of females as verse 27 says, "*So God created man in his own image, in the image of God he created him, male and female he created them.*" This is indicative of the equal importance that God placed on both male and female. Though views vary widely, some Old Earth Creationists believed that God made pre-humans like Neanderthals, and Homo erectus, but they did not have the image of God or spiritual dimension that modern man has. Modern man started with a single Adam and Eve with the image of God implanted in them."

"Dr. Delzer, Dr. Delzer," Kristina was quietly raising her hand and asking.

"Yes Kristina?"

"Is there any hard evidence that mankind came from only a single man and woman?"

"A very interesting question. If you check the mitochondrial DNA of females and the Y Chromosomal DNA of males, it does indicate a single male and female[1]. This is commonly interpreted in secular circles as the woman to whom our Homo sapiens lineage can be traced back to; a bottleneck of one woman in the evolution of man. The same can be said of the Y Chromosomal Adam who was not necessarily married or "joined" to mitochondrial Eve or existing in the same time. However these results can be used to justify or support the argument that humanity came from one man and woman if your worldview will allow it. Look in the web encyclopedia Wikipedia on mitochondrial Eve. They even have a section explaining why the mitochondrial Eve is *not* the biblical Eve. Wikipedia is not known for being very fair to Christianity and other religions. Why do they feel the *need* to explain why the two are not the same? Read their arguments, I don't find them all that compelling. So yes in a nutshell, there is some hard evidence, if you are open to it. It also implies strongly that Neanderthals are not related to humans, and we have some fragments of tissue suitable for DNA testing from Neanderthals, and the results tip the same way[2]. We did not come from them."

"Now let's finish this up with the seventh day of creation. Chapter 2 of Genesis verses 1 through 3. This is in the second chapter of Genesis. This is evidence that the medieval scholars, who added the verse numbers and chapters to the Old Testament, may have got it wrong. This is not the fault of the text, just the medieval scholars who added the chapters around 1200 AD and the verse numbering in 1551 AD. *Thus the heavens and the earth were finished, and all the host of them. And on the seventh day God finished his work which he had done, and he rested on the seventh day from all his work which he had done. So God blessed the seventh day and hallowed it, because*

---

1. Rana, Who was Adam?, 137.
2. Rana, Who was Adam?, 186, 189.

on it God rested from all his work which he had done in creation. Now class what is missing?"

Most of the class responded, *"There was evening and there was morning"*

"Oh, have you been exposed to the day of rest idea?" Delzer exclaimed.

"Yes," "Dr. Hanson told us about it," "We discussed this in class earlier," various students called out.

"Okay, I won't repeat that idea though it lends strong support to viewing days as a long period of time in this section of Genesis. Look at the rest of chapter 2 where the creation of the earth and heavens is repeated, and the order and sequence is quite different. The story is being retold from a very different perspective, and drops a lot of hints that we should not be reading these portions in a flat footed overly literal fashion. That is why I do not get caught up trying to squeeze too much detail and order out of Genesis one. The writer of Genesis, probably Moses, had no problem with using different literary motifs to communicate his points. He is not as sequential or specific as our modern scientific society. He doesn't need to be; and it would not have served him well, *or we* well either. But I still find the correspondence between the general order of creation in Genesis and our modern knowledge, a very compelling feature of the Genesis story. You just can't do this with other ancient creation stories; you can't do it at all. They don't even come close or even discuss the same things. The Biblical creation story is utterly unique, and that is why I read Genesis this way and I am an Old Earth Creationist. Are there any questions class?"

There were a lot of questions, but that is another story as Dr. Delzer's lecture ended.

The scientific graveyard of chance and necessity as the source of life is chemical evolution. Chance implies that random chance provides the variation or information necessary for life, and necessity provides the laws of science that steers chance in the right direction for life. Chemical evolution is the theory that turns a batch of lifeless chemicals into a living working cell, bacteria; the first life. It is a story, a theory that makes the movie *Frankenstein* seem truthful, if you fully understand what it postulates. Both Michael Behe[3] and Stephen Meyer[4] have discussed in their books how chemical evolution was not considered a stumbling block in Darwin's day. That is because they did not understand what a living cell really was. In the mid-1800s a cell was a simple blob of gelatinous material with a wall enclosing it and some sort nucleus in the center; that was it. So coming up with that from some

3. Behe, Darwin's Black Box.
4. Meyer, Signature in the Cell.

muck at the bottom of a pond or ocean was not too difficult to conceive, a sideline to be figured out latter. The idea that life must come from life was a recent one confirmed by the brilliant experiments of Doctor Louis Pasteur in France. One hundred and fifty years later the view that the cell is simple has been turned completely upside down. We now know the simplest cell to be one of the most complicated things on the earth, a repository for information beating the encyclopedia Britannica, and a class of machine that has never been made by man—a self-replicating automaton; or in plain English—a self-replicating machine. What is a self-replicating machine? How does something like this assemble itself by chance and necessity in the very non-industrial early earth?

A self-replicating machine can be described as an autonomous robot that makes copies of itself using the raw materials that it gathers from its environment. Living cells do the same thing but may involve a lot more than is obvious from first inspection. But to understand the magnitude of what must be constructed in making a living cell from raw materials, we ought to examine self-replicators first.

Let's imagine a 3D printer that can reproduce itself. 3D printers are the latest rage in computerized devices. Hook it up to a computer, put a spool of plastic wire into the machine, then go to the computer and call up a program to run that will cause the 3D printer to "print" one of its plastic parts that it uses. Well then, call up a master program to print out all of the plastic parts that the 3D printer needs. Done? Not quite, we have to assemble all of those parts together into the 3D printer, so we need to add an assembly machine onto our 3D printer, a type of programmed robotic arm that will put our new 3D printer-robotic arm assembler together and plug it into a wall socket for power. Oh, we need to use the robot assembler part to plug it into a new computer so that it can be run from a program. We need a master program to oversee all of the part making programs and the storage and assembling program. We also need a way of copying all of these programs and transporting them to the new computer, plugging it in, and starting these new programs running in their proper sequence. We also have to manufacture and assemble the new computer; we can't keep using the old computer and programs on each new 3D printer. Otherwise, it would not be truly self-replicated. So we are going to need to expand that robotic assembler arm into a series of robotic arms and machines capable of putting a small controlling computer together. These robotic arms and machines will have to be copied and built and programmed appropriately from some computer somewhere, maybe our original one that we are going to have to build. Hmm, that means we have to make metal wire, plastic for semiconductor chips, silicon and indium and hydrofluoric acid for etching the semiconductor chips, vacuum pumps for the vacuum technology necessary for the semiconductor chip making, mining and refining technologies and their factories, to produce the metals, hydrofluoric acid, and other exotic chemicals needed for the process. Steel, iron, and copper require blast furnaces, mining machines, and ways of detecting the sources of these necessary raw materials. You don't expect the iron ore to just float up right beside the 3D printer-robot arm-computer-computer manufacturing-self-replicating machine. You have to find it and bring it to the self-replicating machine. That implies some sort of transport mechanism, maybe a dump truck. Plastics are critical for that spool of plastic you insert into the 3D printer, but plastics come from crude oil. Crude oil has to be refined, so it looks like we may have to build and assemble drilling rigs to mine the oil, and refineries to break it down into the necessary raw chemicals, and then processing plants to mix and heat the chemicals properly to make plastic and then assemble into a usable spool of plastic wire. Then we are going to need a robotic arm to plunk that newly manufactured spool of plastic back to the newly created 3D printer-robotic arm assembler-computer factory-mining and drilling machine-oil and plastic processor-steel mill-mining excavator-transporter mechanism machine-and master control computer. Oh, don't forget the program that controls everything and has all of the plans for making and running *everything*, stored in its database. Who wrote that? Or did it just randomly occur? Maybe just turning the machine on and off enough would cause enough random variations in the memory to accidently start and *evolve* this program to its present magnificent state.

Is your head starting to spin? Have I forgotten some critical items? Most likely I have missed many, many other items. I haven't even started talking about that new wall socket into which we have to plug the new working copy. The brilliant Hungarian mathematician John von Neumann was one of the primary architects of the modern electronic computer developed during World War II. He like Michael Polanyi was jokingly called an extraterrestrial, "because anybody that smart cannot possibly be human." He was one of the first to seriously propose how to build a self-replicating machine in his book *Theory of Self Reproducing Automata*[5]. To even come up with a possible design he had to introduce many significant limitations and do it in a fictional two dimensional "automata" space. This might explain an interesting quote made by John von Neumann,

> There probably is a God. Many things are easier to explain if there is than if there isn't.[6]

5. Neuman, Theory of Self Reproducing Automata.
6. Macrae, John Von Neumann, 379.

A materialist could argue that the analogy just described does not accurately describe a self-replicating cell. Living cells do not plug into wall sockets, have hard drives, or need to mine copper ore. In a sense that is true, but in another sense the analogy is more accurate than first suspected. Living cells require power to drive their operation; high energy molecular compounds that are created in the cell from solar energy, or consumed food. They require a means to get their food or raw materials, either locomotion to find and eat food, or a sorting apparatus that distinguishes the raw materials that sweeps by it, that selects what is useful, and what should be rejected. They require a means of building many different parts from the raw materials brought in, assembling those parts, transporting them to different parts of the cell, and starting their operation. They require a means of getting rid of waste, a power plant to drive the cell's machinery, a wall or protection to hold the cell together, and sometimes defense mechanisms to fight off predators. They also need an internal *computer* that *runs* the necessary programs to drive the whole living and replicating processes, and a recording and memory database system that saves the master *program* of life. Let's look at those words:

computer . . . runs	this implies a *machine*
program	this implies *information*

Michael Polanyi was one of the first to recognize what this meant for living organisms. Living organisms are machines. At first glance this proposition seems to be a very materialist perspective. But Polanyi saw the hidden message of this statement and turned this materialist viewpoint on its head. *Reductionism* is the idea that everything can be reduced down to chemical and physical laws. Saying that living things are just very complicated machines is a statement *reducing* their operation down to just the laws and actions of chemistry and physics. Polanyi carefully examined the nature of machines, even the simplest like a wrench, and realized a key feature of machines that made reductionism by analog to machines impossible.

> We have seen that a tool, a machine or a technical process is characterized by an operational principle, which differs altogether from an observational statement. The former if new is a discovery, represents an invention and can be covered by a patent: the latter, if it is new is a discovery which cannot be patented.[7]

What is he talking about here? Operational principles are the rules we use to perform certain functions with certain machines. The simple machines that we construct all machines from are the lever, wheel and axle,

7. Polanyi, Personal Knowledge, 328.

pulley, inclined plane, wedge, and the screw. An inclined plane allows us to raise a heavy weight up a certain distance using less force by directing that force and weight up an inclined plane. That is the operational principle of an inclined plane. A physical and chemical analysis of that inclined plane will tell you it is a pile of dirt shaped somewhat like a shallow triangle, but nothing at all about the fact that Chevrolet Silverado pickup trucks can haul half a ton of dirt 100 feet up that dirt triangle, whereas they could never do it if they tried to climb 100 feet straight up the side of a cliff. You could patent the idea of a dirt inclined plane held together with special cement, but forget about trying to patent the idea of a shallow triangle. A complicated machine is made with many, many simple machines, and thus the operational principles become more complicated and turn into a coherent machine *design* and *purpose*. The design and purpose become patentable under the unique set of operational principles that the machine uses. But no objective scientific investigation of the physio-chemical properties of that machine will ever tell you what it *is*, and *does*, and *why* it does it that way. No strictly physio-chemical investigation will tell you the *design* or *designer*. That has to be inferred from reverse *engineering* principles. As Polanyi eloquently explains:

> The operational principles of machines are therefore rules of rightness, which account only for the successful working of machines but leave their failures entirely unexplained . . . A physical and chemical investigation cannot convey the understanding of a machine as expressed by its operational principles. In fact it can say nothing at all about the way the machine works or ought to work.[8]
>
> This result is crucial. I shall repeat it therefore once more in concrete terms. We have a solid tangible inanimate object before us– let us say a grandfather clock. But we do not know what it is. Then let a team of physicists and chemists inspect the object. Let them be equipped with all the physics and chemistry ever to be known, but let their technological outlook be that of the stone age. Or, if we cannot disregard the practical incompatibility of these two assumptions, let us agree that in their investigations they shall not refer to any operational principles. They will describe the clock precisely in every particular, and in addition, will predict all its possible future configurations. Yet they will never be able to tell us that it is a clock. *The complete knowledge of a machine as an object tells us nothing about it as a machine.*[9]

8. Polanyi, Personal Knowledge, 329.

9. Polanyi, Personal Knowledge, 330.

Machines have a plan, machines have a purpose, and machines have a *designer*. Thus identifying a living cell as a machine militates against the very idea of random non-intelligent construction, unless you have a good cause for redefining the very essence of what a machine is. This is why Polanyi argued over and over that life is not reducible to chemistry and physics, and thus the argument that life can arise spontaneously out a correct grouping of chemicals and input of energy is flawed to the core. But his argument did not stop there. Let us examine the living cell as a self-replicating machine[10]. First let us identify some critical parts of the cell.

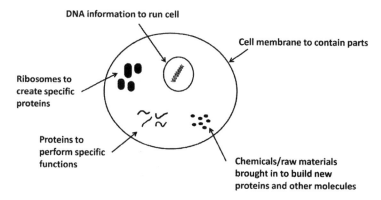

Some critical parts of a living cell

Next the cell starts to make new components much like our self-reproducing 3D printer. The ribosomes take information transferred from our DNA database via a molecule called mRNA and a host of escorting proteins (also called enzymes when used in catalyzing specific chemical reactions), and then use that in conjunction with tRNA molecules to build new proteins out of combinations of 20 different possible amino acids. These proteins are then folded into the exact configurations necessary to make a new protein *tool* or *machine* to accomplish a given task. Some of these new machines are ribosomes, enzymes, and thousands of other critical needed devices to run the cell. Like the 3D printer, you need chemical power to run the reactions (ATP molecules turning into ADP and AMP molecules), chemical conveyor belts to move things, chemical doors and gates to allow access and deny access at the right time, chemical raw material import, chemical manufacturing of tools (proteins/enzymes, and RNA molecules), and of course a master chemical computer and program to run and coordinate everything.

10. Stevens, "Self-Replicating Machines."

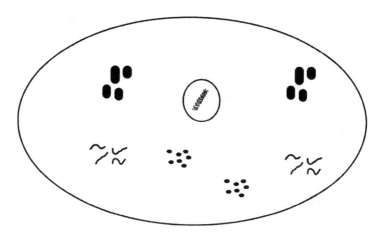

**Cell expands its membrane and creates duplicate
ribosomes, proteins, and other necessary machinery**

Like our previous 3D printer with its robotic arm it is necessary to
find and arrange the parts into the proper places, and then the computing
and control part of our cell (computer??) with its master program must be
duplicated, and then placed in the proper place.

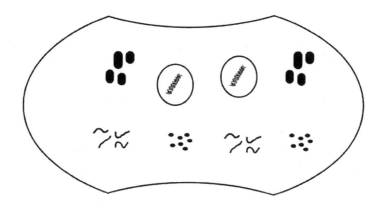

**The DNA is duplicated, parts moved
into proper spatial orientation, and
cell membrane starts to pinch off**

The cell membrane starts to pinch off in the middle under the influence of a special constricting band created by the cell for this job. Finally the two identical halves are separated, and the new duplicate cell is functioning.

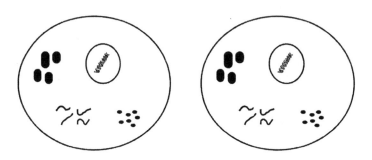

Cell divides into two running living cells

The simplifications used here in comparison to the "known component" of what happens when a cell divides are staggering. An entire book dedicated to the process might get you started. A BS, MS or PhD in biology, and biochemistry "might" give you a good overview of the process, at least what we presently know; which is not much compared to what we really need to know to fully understand it. As a professional chemist it would be easy for me to launch into a detailed technical description, but the technical jargon, needed background in chemistry and biology, and flood of details will overwhelm the non-specialist reader and likely cause you to miss the point in a flood of eye-glazing details. To the specialist I apologize, but the object is to make this book approachable to a wide variety of readers so I will be trying some different approaches to communicate the soul of a very technical and sometimes difficult subject.

The first thing you will notice in this process is that, like the self-reproduction of the 3D printer, incredible quantities of information are involved. Construction blueprints on what parts go where, and in what order. How do you assemble these parts, where do you place them, and then how do you get them running and doing their job? Scientists have identified that DNA is the primary information carrying molecule in a cell. It is possible that much, maybe all (??) of the program, used to build and run a cell is stored in the DNA strand that lies at the heart of every cell. This information is very specified information. It begs the question of how did it get there and where did it come from. The specific structure of DNA was figured out by James Watson and Francis Crick in 1953 at Cambridge University. It took another 10 to 15 years to understand how that information was coded into the DNA strand and then translated and transported to the ribosomes to produce new protein machines. By the mid 1960's Polanyi realized that this information was independent of the DNA strand, and that in fact it had to be. The DNA code could be recorded on paper, magnetic tape, computer hard drives, or books like any other kind of specified information. In fact the human genome project was completed in 2003 with a complete mapping of a typical human genome. This information was extrinsic and independent of the medium that carried it (DNA strand), and completely unexplainable by any known chemical or physical phenomena, implying that the living cell could never be *reduced* or explained by scientific laws. Polanyi's argument was persuasive enough to be published in the very prominent journal *Science*[11].

From a theist's perspective the source of the information is obvious, God. Based on our common and historical human experience, only an intelligent agent can communicate specific information. A materialist will have to postulate that information can arise from non-intelligent sources, and this can be the source of a lot of anxiety, if you take Polanyi's arguments seriously.

Mary, Charlie, Kristina, and Ellie were lounging in a deserted faculty meeting room with their homework and study notes scattered sporadically about the table around which they sat. The topic had wandered from their studies to the origin of life lectures that Mary, Kristina, and Ellie had attended. Confusion and frustration was in the air, tempered by occasional bites of pizza, and sugar dusted donuts.

"Okay, so chemical evolution is the greatest problem to explaining life that a materialist faces, but why isn't this better known," Ellie was puzzled. "Where are the debates discussing this great quagmire in the university?"

Charlie replied, "Good point, but if the modern university is leaning in a materialist direction why advertise an area of research that has eluded explanation for the last 65 years. To follow the arguments you really need courses in general chemistry, organic chemistry, biochemistry. Some courses in cellular biology and geochemistry would not hurt. Three reactions into the debate the popular audience's eyes glaze over and the debaters are basically talking to themselves. Tomorrow's paper reads, "Materialists cower before creationists after trying to explain away the Concerto effect." Just isn't going to happen. People need something approachable that they can follow before they are going to hear both sides."

"Do we really have to get that technical, to get the points across?' Ellie gestured with her hands in Charlie's general direction. "I haven't taken

11. Polanyi, "Life's Irreducible Structure." 1308–12.

much chemistry, but I am college educated; well almost. If you really know a field you ought to be able to get the main points across without getting lost in technical details. I sat in on a metallurgy seminar where the professor brought us up to speed on the latest advances in a day, and I am not even a science major."

"How did he do it?" asked the rest.

"You mean she; she just skipped the math, and used simple analogies to build up the specialist's vocabulary. I'll never be able to do work or advance the field but I do get the point of why they are doing what they do."

"I think I can do that for you here, Ellie," Mary was smiling, "At least I would like to try; Charles will you help me?"

"Okay," replied Charles, Ellie and Kristina.

"Actually Mary, a well-known agnostic chemistry professor has written such a book."

"Oh, who is it Charles?" queried Mary.

"*Origins: A Skeptic's Guide to the Creation of Life on Earth*, by Robert Shapiro[12]. He still throws out his own theory of how life could spontaneously arise, but he completely trashes all the other theories and is honest about the problems he is facing."

"Well educate me then!" Ellie replied.

Mary strolled over to the dry erase board on the wall and drew a bunch of circle atoms with bonds. "Let's pretend that atoms are like tinker toys balls with bonds attached. Here are the possible arrangements of carbon.

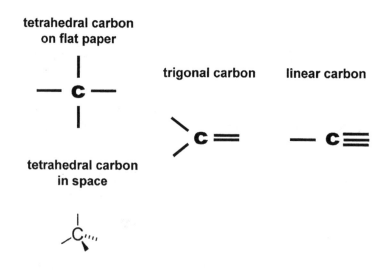

12. Shapiro, Origins.

Here are the possible arrangements of nitrogen.

nitrogen on flat paper

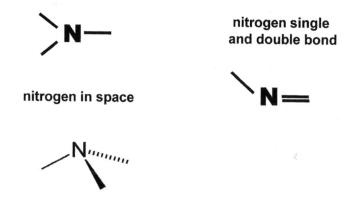

nitrogen single
and double bond

nitrogen in space

Here they are for oxygen,

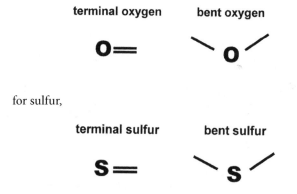

terminal oxygen   bent oxygen

for sulfur,

terminal sulfur   bent sulfur

and lastly here is the only arrangement for hydrogen.

hydrogen

Now let's hook these together to make molecules. Here is how we can hook carbons together to make a hydrocarbon, or alkane. First let's start with methane.

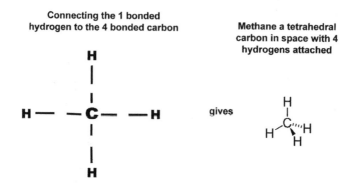

Connecting the 1 bonded hydrogen to the 4 bonded carbon

Methane a tetrahedral carbon in space with 4 hydrogens attached

gives

Now let' put together a two carbon molecule called ethane.

Connecting 1 bonded hydrogens to two 4 bonded carbons

Gives ethane

Now in the diagram following are several different ways chemists have of representing or drawing methane, ethane, and the other "alkanes" (a class of molecules composed of carbons and hydrogens bonded with only single bonds). We have the formula CH_4, or C_5H_{12}, etc. We have the ball and stick representation where the balls represent atoms, and the sticks the bonds connecting them. Next you will see the stick representation where the carbon atoms are represented by the darker intersection or joint where the bonds all meet. Moving to the right you see the space filling model where the atoms are represented by spheres and when two spheres overlap, you know there must be a bond between those two atoms, but you cannot tell what type of bond it is (single, double, or triple). But you get a better picture of how much space each atom occupies—hence the name space filling.

Notice on the diagram following, on the far right, is the wire frame representation. We can simplify the drawings even more by not including the letter C for carbon. A carbon atom is assumed to be at the end of every short line, and everywhere two short lines meet. If the atom is not carbon, but nitrogen, oxygen, sulfur or some other atom then you put the letter in at the end of the short line or at the junction of two lines. Since we can guess at the correct number of hydrogen on each atom from how it is bonded to the other carbon, nitrogen, oxygen, etc. atoms, then we don't draw them in, they are assumed.

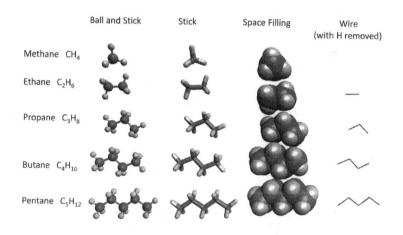

Here's how you can make alkenes.

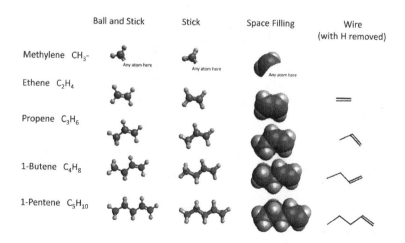

Notice that the carbon-carbon double bond does not show in the stick or space filling view. Remember also that parts of the molecule can rotate around a single bond, but double bonds do not rotate, thus the molecule cannot rotate or change orientations around a double bond. The 1 in front of the name indicates the position of the double bond. For example in 1-pentene the double bond is on the extreme right. But flip the molecule left to right puts the double bond in the extreme left carbon-carbon bond position, so the first and fourth position are the same. Moving the double bond to the second or third position is the same also or 2-pentene, but that is getting into more detail than we need." Mary paused a second.

"Please explain what you mean by parts of the molecule rotate around a single carbon-carbon bond but not a double bond?" queried Ellie.

Mary then carefully drew another diagram, "Here is how an alkane can rotate about its C—C bonds to form different conformers. The chemical formula C_5H_{12} is the same for all of the conformers, the bonds all go to the same atoms, it's just that the molecule has twisted to a new position or structure."

Pentane conformers – molecule changes shape by rotating around C – C bonds

Pentane C_5H_{12}	Ball and Stick	Stick	Space Filling	Wire (with H removed)
Conformer 1				
Conformer 2				
Conformer 3				
Conformer 4				

"What is interesting," interjected Charles, "Is how the long chain of atoms acts as a flexible linkage, a series of axles, joints, cranes, and arms, the movable part of a molecular machine."

Mary looked puzzled, "What do you mean?"

"I'll explain later."

Mary continued, "Well, methane, ethane, propane, butane, pentane, hexane, heptane, octane, nonane, decane, etc., represent a series of molecules we call alkanes, or aliphatic hydrocarbons. We get these from refining crude oil and they help make up the mixture we call gasoline, that our cars burn. They tend to be oily and the more carbons you add the heavier and more oily or grease-like they become. They do not dissolve in water. And they don't do much—that is they do not react with many compounds to form new compounds. To get these alkanes, alkenes, amines, and other types of molecules to react we have to add nitrogen, sulfur, and oxygen atoms in certain atomic groupings we call functional groups. These functional groups are what react and give us the field of organic chemistry."

"Why would I want to know this? How does this fit in with understanding living things," Ellie was puzzled.

"Well, the chemistry and reactions that take place in living things is basically organic chemistry, the chemistry of carbon, hydrogen, nitrogen, oxygen, and sulfur, etc., where we put the chemicals in a flask with some solvents and heat them up to make them react. Living things run these same type of reactions in the warm water of a cell, using enzymes to catalyze (that is, make them react faster) the same types of reactions," Mary quickly answered.

"Oh . . . ," now Ellie got it. "What are these functional groups?"

Mary went to the dry erase board again, and carefully drew the following, "This is how you can put nitrogen in an alkane to make an amine."

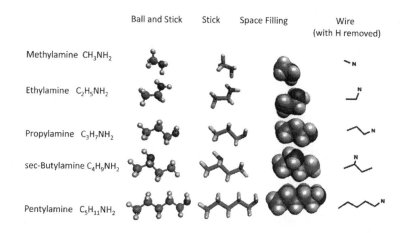

Kristina spoke up, "Hey what is going on with the sec-butylamine. The nitrogen (darker atom) is not on an end carbon atom, but the second from the left carbon atom."

"That is exactly right!" exclaimed Mary, "Hence the sec- in front of the name butylamine. You can put the amine functional group any place as long as you have only three groups attached to its three bonds. For example; look at this compound drawn in wire format."

"Whoa!" spat out Ellie, "If you moved the three arms around you could make the three amino groups on the ends wrap around something like a big atom."

"Ellie, you are becoming a genius chemist already. Those three carbon arms do bend and flex, and allow the amino groups on their end to wrap around atoms or other molecules they are attracted to. Chemist have a name for that, it is called chelation." Charles had jumped back into the conversation. "You know, I have never thought about molecules in quite that light before, we are making little molecular machines. Little molecular machines designed and created by intelligent chemists to do specific tasks. There is something profound in this . . . moveable parts, linkages or joints, action causing functional groups . . ." Charles was quickly lapsing into an internal lecture of his own making.

Mary was not about to wait that long on Charles.

"Okay, now let me show the basic functional groups, there are others, but these are some critical ones. I am going to use a capital R to represent any old groups of atoms, or a smaller subset of a molecule. Here is how I could represent methylamine, ethylamine, propylamine, butylamine, etc.

$$R\text{—}NH_2$$

I don't know what R is, but I can clearly see that it is attached to an amine functional group at the end."

"Mary, could you generically represent 2-butylamine?"

"Good question, here is how, but it could mean other similar molecules also."

$$R\text{—}NH\text{—}R$$

"Does it make sense now?" queried Mary.

"Yes it does," replied Ellie.

"Here are the basic functional groups that make things happen in organic chemistry and biochemistry."

"What is a functional group?"

Mary paused a moment and thought, "It's the small grouping of carbon, oxygen, nitrogen, or hydrogen atoms in a molecular sub-group that make a molecule react and do things."

"Oh, that's right, you mentioned that earlier. But what sort of things do functional groups do?" Ellie reddened a little.

"Ah, they cut or cleave certain molecular bonds in other molecules, they pull hydrogens off other molecules, they pull OH groups off, they attach two molecules together, they push hydrogens onto other molecules, they turn single bonds into double bonds by removing hydrogens, and . . ."

At this point Charles jumped in and said, "They act like molecular scissors; that cut molecules into pieces; or like welders that join two molecular pieces together; or like pinchers that squeeze parts of molecules off; or glue and scissors that cut off part of a molecule and put another part on. They are the parts that *do* chemistry, make things happen, and cause molecular change. Do you get it Katrina and Ellie?" They both nodded agreeably, but were not so sure to themselves.

"Wow, little molecular machines with joints, arms, extensions and cutters, welding machines, and bite-off clamps. I never thought of organic chemistry from a machine viewpoint before." Charles was getting excited.

"Calm down, Charles!" Mary was trying not to roll her eyeballs, "Let's go over some of the basic functional groups we often see. Each one has a unique name."

This is how you can put oxygen in to make an ether.

Now this is an alcohol.

This is a ketone.

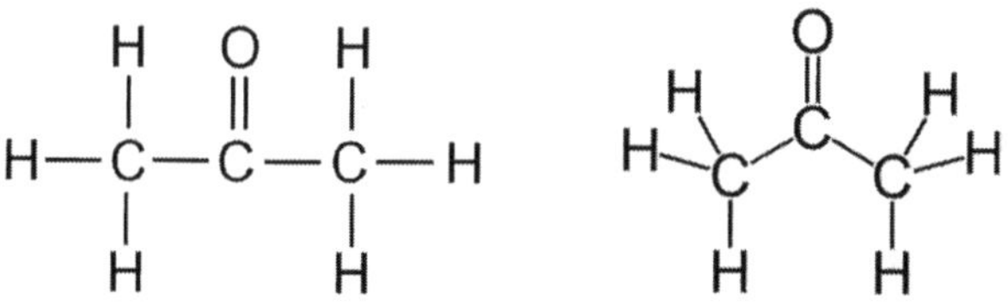

$$H-\overset{\overset{\displaystyle H}{|}}{\underset{\underset{\displaystyle H}{|}}{C}}-\overset{\overset{\displaystyle O}{||}}{C}-\overset{\overset{\displaystyle H}{|}}{\underset{\underset{\displaystyle H}{|}}{C}}-H$$

This is an ester.

$$H_3C-\overset{\overset{\displaystyle O}{||}}{C}-O-CH_3$$

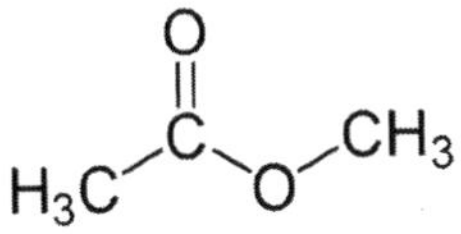

Here is another very important functional group, an organic acid.

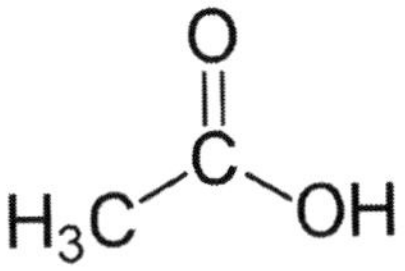

$$H_3C-\overset{\overset{\displaystyle O}{||}}{C}-OH$$

This functional group is reactive and is present on every amino acid molecule. It will react with an amine group on another molecule to bond the two molecules together. An organic acid bonded to an amine makes a new functional group called an amide. This amide linkage is what holds amino acids together to make a protein. The reaction looks like this.

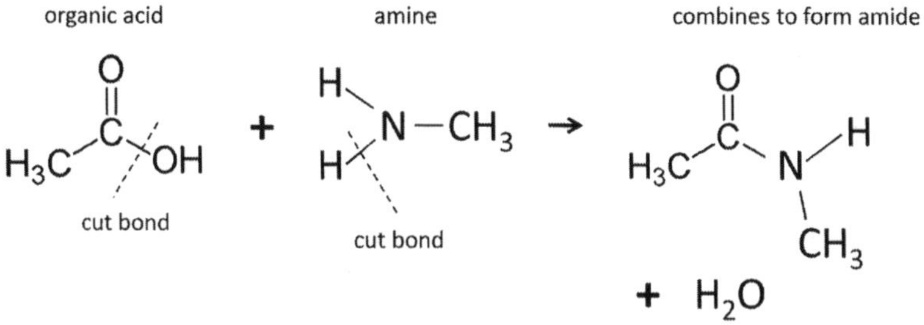

organic acid   amine   combines to form amide

$$H_3C-\overset{\overset{\displaystyle O}{||}}{C}-OH$$

cut bond

$+$

$$\overset{H}{\underset{H}{>}}N-CH_3$$

cut bond

$\rightarrow$

$$H_3C-\overset{\overset{\displaystyle O}{||}}{C}-N\overset{-H}{\underset{|}{\underset{CH_3}{}}}$$

$+\ H_2O$

The cells in your body use this type of reaction to make the proteins needed for constructing your body. Now let's put six carbon atoms together is a circle, to make cyclohexane.

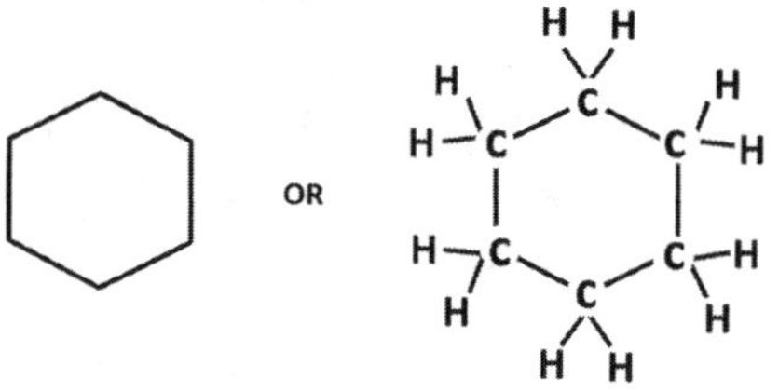

Now let's put some double bonds in to make 1,3,5-cyclohextriene.

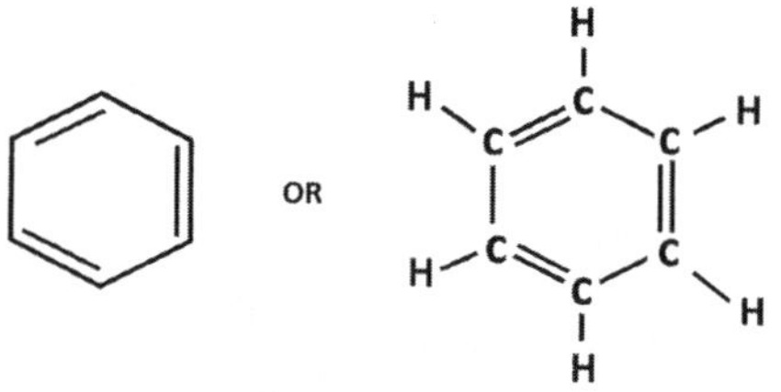

Now average the double bonds over the circle to give you 1½ bond between each carbon atom. This is benzene and it is an aromatic molecule. Aromatic molecules are a very special class of molecules in chemistry. You need double and single carbon—carbon bonds alternating in the molecule. If created properly the double bonds will smear out over the whole alternate set of carbon—carbon bonds such that it averages to 1½ bonds between the carbon atoms. That happens only in ring molecules like benzene, etc., and we represent this smearing of the double bond around the ring with an inner circle as shown in the benzene molecule in the next figure.

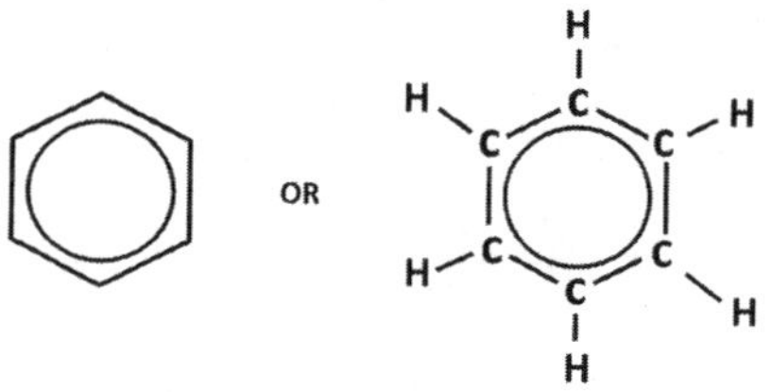

There is an important property of aromatic molecules that needs discussion. The partial double bond around the ring prevents any of the carbon—carbon bonds from rotating. Thus aromatic molecules are rigid and flat. They have only one conformer, no other is possible unless you attach flexible movable groups of atoms around the outside edge of the ring. Thus aromatic molecules can be flat stable platforms to which we attach other flexible moving groups of atoms to—to do things—to do chemistry. Then aromatic molecules can have different conformers or shapes. Thus aromatic molecules make great platforms for attaching functional groups to at fixed points; and at the correct spacing. They are the platforms for our molecular machines. Now let's put two benzenes together to make another type of aromatic molecule called naphthalene.

It makes a great rigid platform also. Now let's make a phosphate.

We are going to use this group to attach other molecules together to make DNA. Now here is a molecule that we call a purine. It is aromatic like benzene and naphthalene and likewise a rigid platform imbedded with nitrogens that have a bit of excess negative charge on them.

Here is another single ring aromatic molecule called pyrimidine. Like benzene but with imbedded nitrogens that carry a bit of excess negative charge that they stole from the rest of the molecule.

Sometimes purines and pyrimidines can be attracted to each other through a very weak type of chemical bond called a hydrogen bond. It looks like this. The dotted line indicates the very weak H—bond or hydrogen bond. Notice how the embedded nitrogen with a slight negative charge is drawn to atoms that have a slight positive charge on them. The hydrogens attached to the nitrogen have electronic charge stolen from them by the nitrogen making them slightly positive.

This type of bond occurs only between N-H and N or O or S, or between O-H and O or N or S, etc. Here are some examples.

This type of very weak bond holds a lot of biochemical structures together and is responsible for life being possible. It makes a great chemical "zipper" or "sticky note" type of glue to hold certain molecules together so that they can be removed and reattached repeatedly."

"Charles anything you want to add at this point?" Mary asked.
"One year of organic chemistry on one white board, very impressive!"

"Oh, shut up Charles," Mary intoned while grinning. "Are you following me Ellie?"

"Yes, I think so. The sticks on the sides of the atoms represent bonds; put two atoms together by their sticks and you have one bond between them. If there are two sticks close together then that represents a double bond. You can't put a single stick projection together with a double one, just another single stick projection, right?

"Yes."

"Kristina, do you get it?"

Kristina looked away from the board and replied, "Yeah, just keep on going."

Mary approached the dry erase board again and erased the previous diagrams. "Now let's draw some important biochemical structures. Here is type of molecule called an amino acid. The letter R represents any grouping of atoms. R varies from amino acid type to amino acid type.

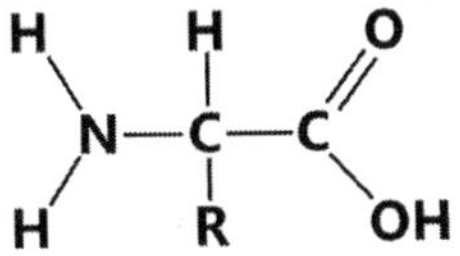

There are 20 different amino acids that make the proteins in most living cells and organisms, and here they are.

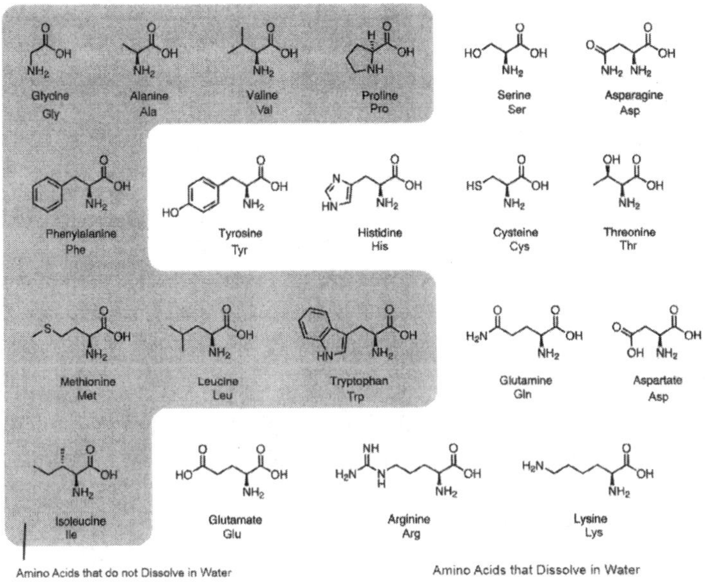

Amino Acids that do not Dissolve in Water   Amino Acids that Dissolve in Water

Notice how each of the 20 amino acids has a three letter abbreviation. Now you can chemically bond these amino acids together to form a peptide, or if it is long enough (50 amino acids) we call it a protein. We can use these three letter abbreviations to specify how the amino acids are connected together. Here are three glycine amino acids bonded together to make a small peptide called triglycine. The order of these amino acids in this peptide is called the primary sequence and is written like this in abbreviated form: Gly-Gly-Gly. The actual chemical structure would look like this.

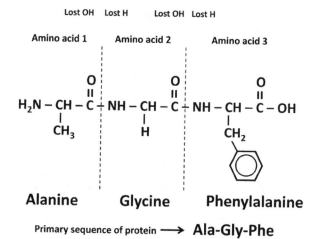

See how the middle glycine molecule has lost an OH and an H in combining with the other two glycine amino acids and the glycine on the right has lost an OH and the glycine on the left has lost an H atom. For two amino acid bonds formed, two molecules of HOH or H2O or water were produced. This also plays a role in chemical evolution to be discussed later.

I can change the three amino acids to any of the 20 basic amino acids that I want, like this example.

In fact I can easily calculate how many possibilities of peptides I can get with just three amino acids bonded together. That is 20 x 20 x 20 = 20^3 = 8000 possibilities. Now let's try it with just 10 amino acids."

Charles pulling out his calculator punched in the numbers. "That is 20^{10}. I got 1.024 x 10^{13} or 1024000000000 possibilities. Hey try 80 amino acids; that is the length of a typical small protein in the human body."

Mary looked at Charles with a jolt, "Ah, yes."

"My calculator overflowed, it cannot hold such a large number." Charles looked up at Ellie. "This is an example of combinatorial inflation that Dr. Delzer was talking about. The number of possible proteins goes out of sight to numbers beyond the number of atoms in the universe or possible reactions you could do per second in the entire lifetime of the universe if you make your protein long enough."

Mary replied, "What are you getting at Charles?"

"Suppose in the early earth we have to randomly assemble a working protein in the primordial soup of the early earth, where life supposedly formed. The correct protein sequence that gives you what you need to get a replicating molecule going or something that helps kick start life is going to require a fairly large protein, at least. The odds of stumbling on the right sequence by chance are astronomically against you."

"Yes, whatever. Can we get back to my biochemistry overview, Charles?" Mary was a bit peeved at the interruption. "Now as Charles has so kindly pointed out, the primary sequence of a peptide, or a protein, is determined by the order of the amino acids in the chain. But that does not guarantee that the protein will have any useful biological function at all. It must have the right conformer or folded structure. See here are some different conformers of a protein and here is a big folded protein that represents one possible conformer of that protein. The particular folding or conformer that a protein adopts in its biologically useful form is called the secondary structure and tertiary structure of that protein." Mary drew them all on the board.

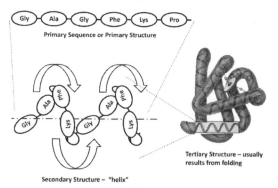

Primary Sequence or Primary Structure

Secondary Structure – "helix"

Tertiary Structure – usually results from folding

"Do proteins flip back and forth between all of their possible conformers or folded structures, and how many folded structures or conformers do they have?" asked Kristina.

"It depends on their size or length and how much thermal energy or heat is in their environment."

"Huh?"

"Small proteins (peptides) have fewer conformations and usually just one that they usually assume or fold to, if they fold back on themselves. They usually do it automatically in a solution of water. Scientists would call that the lowest energy structure of that protein," answered a smug looking Mary.

"What about larger proteins, like the ones we see in living cells?"

Charles cut in, "That's a whole different story, part of the cutting edge of research in protein science."

"Go ahead and take it Charles," said Mary.

Charles taking the cue started in, "Larger proteins have so many conformers, so many possible folded structures, that it is often impossible to identify the lowest energy structure of the protein, and we don't even know if the lowest energy structure is the one that will give you the biochemical function that you want. Imagine that a long protein strand is like a sticky flexible cooked spaghetti noodle. You pull it out of the pot and throw it up in the air, and let it land on the table. How many different ways will it fold up and stick together in a folded mess on your table, lots. Of those gadzillion ways of folding only one is the correct one that will allow that protein to perform some biochemical reaction, like opening up and dividing a DNA segment, or cutting a protein or RNA strand at a specific point in the chain."

"But how does the protein chain get folded into the correct shape in that case?" asked Ellie.

"That is what the ribosome in the cell does. Messenger RNA (mRNA) from the nucleus of the cell was copied from the DNA stored there. The DNA has the blueprint, the primary sequence of the protein coded into it. There upon proper initiation signals (more proteins and enzymes) a certain portion of the DNA is copied to a synthesized strand of mRNA and then the mRNA is escorted by various proteins over to the ribosome in the cell where it is attached and feed into the ribosome much like a magnetic tape or flash drive into a computer printer. The code is read and the protein constructed with the correct primary sequence. As it is being constructed it is folded into the correct structure inside the ribosome, before it is released into the cell to do its job. At least that is what we think happens." Charles sat down with a satisfied but puzzled look on his face.

"How do the ribosomes in the cell do this?"

"We don't know."

"How does the ribosome know which conformer to fold it into? That is, how did the correct folding structure of the protein get put in the ribosome and how?

"We don't know."

"Isn't this information; very specified information of some sort?"

"Yes," answered a puzzled looking Charles.

"How many possible conformations or folded structures does a large protein have?" asked Kristina again.

"Whoa, that is not an easy question, but if we assume along the chain there are at least two rotatable bonds per amino acid with at least 2 to 4, let's say 3 for argument's sake, rotatable positions that the different chemical groups can assume, then for a tripeptide that gives around 3 x 3 x 3 x 3 x 3 x 3 = 3^6 = 729 possible secondary structures give or take 50% or more," Charles was busily calculating to the amazement and awe of Ellie and Kristina.

"How many possibilities are there for a protein that is 80 amino acids long?"

Rapidly punching the numbers into his calculator Charles looked up and smiled.

"Kristina that will be $3^{2 \times \# \text{ amino acids}}$, or 3^{160} or 2.185 x 10^{76}, or 2185 with 73 zeroes after it. My calculator almost overflowed on that one."

Ellie commented, "It is combinatorial inflation again, the number of possible conformations is out of sight, almost unbelievable."

Charles started again, "You know what that means, for chemical evolution to randomly hit upon the right protein, in some early primordial soup scenario is the probability of getting the right primary sequence, and then folding into the right conformer (without the ribosome, which does not exist yet) or folded structure before the molecule does anything useful. It is combinatorial inflation times combinatorial inflation. There isn't enough time in the universe to pull that one off. But this assumes all conformations are equally probable, which is not strictly correct, but it still does not change the probabilities that much.

"How many of those quazillions of possible conformations of our 80 amino acid long protein have some kind of useful biological activity?" queried Kristina.

Suddenly the side door sprang open, shoving Ellie up against the wall. Her soda went flying into the air with a shriek of surprise and horror from the pinned lass. Kristina was wide eyed in shock, as Mary joined in the common shrieks from the frightened company. Charles whirled around to see . . . .

"Lots of them I bet. Lots of those conformations or structures will have biological activity," exclaimed Jon. There he stood in all of his inquisitive wonder, gazing confidently at the frightened quivering mass of humanity trying to compose themselves in the room.

"What in the He . . . heck are you doing? You scared me out of my wits!" exclaimed Mary. "What were you doing behind that door?! Why did you barge in here like that? Are you nuts or just plain rude?!"

"Oh, did I scare you or something?" mumbled Jon.

"Scared! What do you think?! Look at Ellie!"

Ellie was at the moment gurgling quietly and crawling across the floor towards a chair.

Jon realizing his disaster was quick to spring to her aid and help her up with profuse apologies. Charles and Jon got her seated in a chair and retrieved what was left of her soda.

Kristina was impressed. Jon could be a total klutz, but he did know how to apologize. It was obvious his concern for Ellie was not faked. He was genuinely distressed that he frightened her so.

"I happened to be studying next door and overheard your conversation. I was lying low just to see what you would say. But I could not stand not being in on the conversation. I should have given you some warning earlier that I was there and asked to join. I just forgot myself, I am so sorry," Jon was not in his typical know-it-all mood. It was quite different.

"Okay, okay, you have apologized enough," said Mary. "And you Ellie?" Ellie nodded her head in silent assent. "What was your question again?"

"Ah, how many of those quadzillion of conformations of your 80 amino acid protein will have some biological activity?" asked Jon.

Charles intoned quietly, "That is a very good question; no one really knows the exact number. But Douglas Axe, the protein researcher formerly at Cambridge University, performed some very key experiments that estimated the prevalence of biologically active variants or slight modifications (mutants) to the primary sequence of a common biologically active protein. These variations (mutants) can often affect the conformers or folded structures of a protein. He published his results[13,14] in the *Journal of Molecular Biology*, one of the key journals in the field. The percentage he got was very, very low. No, not low, but so low as to be scarce beyond belief. Almost all of the possible primary sequence mutants and conformers and folded structures are *not* biologically active, only a very few select ones. It's combinatorial inflation all over again, and it has huge ramifications for the theory of evolution."

"Really?" said Jon.

13. Axe, "Extreme functional sensitivity", 585–95.
14. Axe, "Estimating the prevalence", 1295–315.

"Really!" answered Charles, "But that is another story that we'll have to discuss later."

Let us return to a brief discussion of biochemistry. I will leave Charles, Ellie, Mary, Kristina, and Jon and pick up the lesson here. Below are a group of molecules called sugar molecules. Ribose is a six membered ring sugar molecule that plays a crucial role in DNA and RNA.

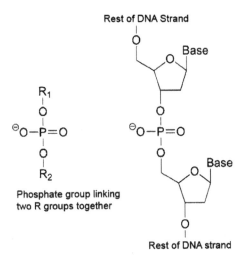

Here is a phosphate linking group. Notice how you can use it to link two other molecules together.

Now connect ribose sugar molecules together with phosphate linking groups to make a long chain molecule or a *polymer*. Notice how DNA uses deoxyribose instead of ribose.

RNA Strand   DNA Strand – one side of DNA double helix

Now each ribose sugar group on the chain can have another molecule attached to it, like a purine type of molecule or a pyrimidine type of molecule.

There are two types of pyrimidine and purine molecules that are found in biological systems. We call them the bases, the DNA and RNA bases; adenine, thymine, guanine, cytosine. How they are ordered in the DNA and RNA molecule is what gives us the genetic code, or program or life.

Adenine — Thymine　　　Guanine — Cytosine

The bases have a tendency to hydrogen bond together like this. Notice how well the purine type of molecule "fits" or hydrogen bonds to its respective pyrimidine type of molecule.

adenine (A)　　guanine (G)　　cytosine (C)　　thymine (T)

Notice that adenine (often abbreviated A) always pairs with thymine (often abbreviated T) and guanine (G) always pairs with cytosine (C). Thus A—T and G—C are the DNA base "pairs." This pairing happens in the DNA molecule. So take a phosphate ribose phosphate ribose etc. polymer and attach one of these four base pairs to each of the ribose sugars groups in the polymer chain and you "almost" have a RNA molecule or strand. Then take another strand or polymer of "almost" RNA and orient it such that the bases on each strand can hydrogen bond to the bases on the other "almost" RNA strand. This will only work nicely if the adenines (A) are hydrogen bonded to thymine (T) and not to guanine or cytosine. And likewise the guanine (G) must be hydrogen bonded to cytosine (C), and not adenine or thymine. The order of bases on the second strand of "almost" RNA must be complementary to the order of bases on the first strand, where adenine (A) is across from and hydrogen bonded to thymine (T) and cytosine (C) is across from and hydrogen bonded to guanine (G). An arrangement like this is the DNA molecule and it only works if the double strands of deoxygenated RNA hydrogen bonded together are twisted together in the classic double helix form of DNA.

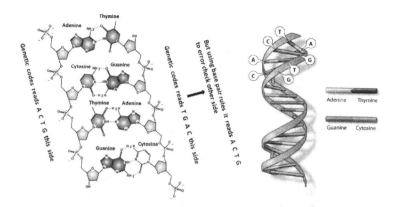

Now the beauty of this system is that you read the code by reading out which base is on one side of a DNA strand. For example: adenine, cytosine, thymine, guanine or ACTG if we read the left DNA strand (left picture) from top to bottom in the above picture. But on the other DNA strand reading the same section you would get the complementary bases or TGAC. Since you know G always pairs with C and A always pairs with T, then you have two copies of the genetic code written into the DNA strand. If you have one side you have ACTG, and if you have the other side you have TGAC which you can decode back to ACTG using the pairing rules. So the genetic code is coded redundantly. This allows for cellular error checking of the code like the best computer information storage systems of today. How interesting. Funny how we are using the word *system* in describing the inner workings of a living cell; like an information system, processing system, construction system, sounds like something constructed by an intelligent entity.

Now there is one modification that needs to be made. In biological systems when you see RNA the base molecule uracil is used instead of thymine. Uracil hydrogen bonds to adenine much like thymine does.

Base Adenine on
DNA half strand

Base Uracil on
RNA strand

DNA half strand

RNA strand

So a GAT section of DNA would be coded as GAU in RNA.

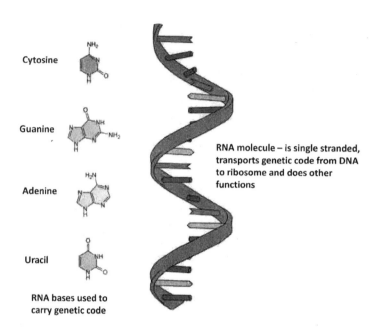

Cytosine

Guanine

Adenine

Uracil

RNA bases used to
carry genetic code

RNA molecule – is single stranded,
transports genetic code from DNA
to ribosome and does other
functions

Here are some very relevant properties of DNA and RNA. DNA is very stable. You can extract it from a cell in a solution of water and store it for months at room temperature on the shelf. RNA is not stable. It reacts very quickly, must be frozen or cooled, and used as soon as possible when warmed up or else it will rapidly decay. It survives best inside a functioning walled cell. This is important to remember when we talk about chemical evolution later.

There are untold numbers of chemical processing systems in a living cell. Some are known, many are yet to be discovered. But the system we have studied the longest, and know something about it is the genetic code translation to proteins, and the DNA replication system. Once the structure of DNA was figured out it took another decade or so of research to discover how the primary sequence of the proteins used in living cells is stored there. Next we discovered how that code was read, copied, transferred to the cell ribosome, and then used to manufacture a new protein. This was first called the "Central Dogma" of Biochemistry. Let us go over it.

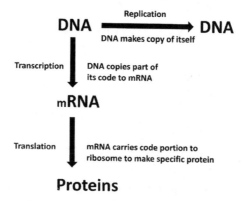

DNA ⟶ **DNA**
Replication
DNA makes copy of itself

Transcription | DNA copies part of its code to mRNA

mRNA

Translation | mRNA carries code portion to ribosome to make specific protein

Proteins

First look at DNA replicating to another DNA strand at the top going horizontally in the above diagram. In the cell, the DNA is tightly winded up in tight coils two or three times (coiled up, then that strand of coils is coiled up again) around globular proteins called histones spaced at periodic points along the DNA strand. Then the histones units come together bunching up the super coiled DNA into a long fuzzy looking strand which is super coiled several more times until the whole incredibly long chain of DNA is tightly packed into a unit called a chromosome, and then stored in the nucleus of a cell. Some simple cells do not have a nucleus but the chromosomes would be somewhere in the middle of the cell. To replicate you would need to first find a starting point in the DNA strand, unwind the super, super coiled DNA until you can get to the portion of the DNA you wish to duplicate (or read). Then you would need to insert that starting point of the DNA strand into a complicated machine made of many specialized proteins that hydrogen bond together perfectly. The DNA strand goes in one end of this protein complex where inside the DNA is split into two strands with their bases attached. Each strand is copied with the complementary base strand which is then attached to the original strand to form another copy of the original DNA.

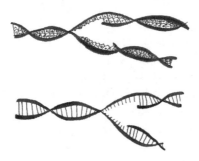

A more complicated diagram showing some of the mind blowing protein machinery used is shown below.

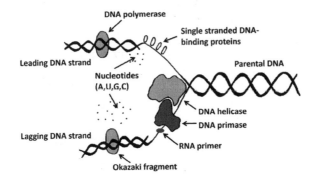

Moving on to the rest of the "Central Dogma" diagram we will look at transcription; how you decode a tiny section of DNA and get the primary sequence of a specialized protein from it.

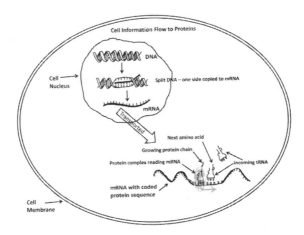

The transcription process of decoding a specific segment of DNA to produce a specific protein varies some from cells with a nucleus (eukaryote) and cells without them (prokaryote); but not that much. Look at the above diagram in the cell nucleus. First the necessary segment of DNA is found and the opposing strands are separated along the hydrogen bonded base pairs. Then one strand is replicated with a piece of complementary mRNA (little m means messenger as in messenger RNA). Many pieces of interacting protein machinery are needed to produce this RNA code segment duplication [duplicated "tape" of the specific information (DNA code)]. The finished mature piece of mRNA is then "escorted" [transfer truck transported out] out of the cell nucleus by other specific proteins through the cell to a piece of cellular protein machinery called the ribosome [CD manufacturing facility]. There at the cell manufacturing facility (ribosome) the DNA code duplicated on the mRNA strand is read and decoded [computer controlled machine lathe or vertical mill] using tRNA (t means transfer as in transfer RNA) to produce a specific protein by reading the DNA code and adding a specific amino acid to the growing chain of amino acids. When the RNA code is read the protein [CD, toaster, scissors] is finished, folded, and then moved out of the ribosome for use in the cell. In the previous paragraph, I used [ ] brackets to denote a very crude machine analogue to what that particular molecule was doing. In doing so, I hope to reinforce the idea that the living cell as a self-replicating machine is not as far removed from the 3D printer example as one might initially think. In technical detail it is actually far, far, more complex.

Decoding the RNA strand for a specific amino acid gets us to the heart of the genetic code. Look at the diagram below. Notice how a tRNA molecule (short T shaped curved segment of RNA has three RNA bases attached at the base of the T (see the three small blocks/teeth protruding in the diagram). The RNA bases match up with the complementary set of three bases on the RNA strand. Each set of *three* RNA bases in a row codes for a specific amino acid which the specific tRNA has attached at the top (represented as a black dot on the diagram below.

Using tRNA to Decode mRNA to Make Proteins

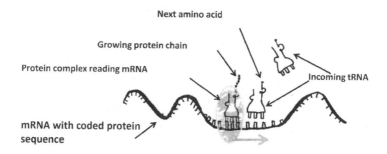

Next amino acid

Growing protein chain

Protein complex reading mRNA

Incoming tRNA

mRNA with coded protein sequence

Thus each of the 20 amino acids has a specific tRNA that has a specific set of three base pairs in a row that codes for it. These three base pairs in a row are called a codon. There is a codon code for each amino acid, for starting the reading of the codons, and for stopping the reading of the codons.

The particular three base pairs that make up each of these codons is the *genetic code*. The chart below shows how the RNA base pairs code for specific amino acids. Remember that RNA uses uracil instead of thymine as a base pair so the base pairing is G to C and A to U, rather than G to C and A to T in the DNA strand. Thus to decode from DNA simply substitute T (thymine used in DNA) for U (uracil used in RNA).

Genetic Code using RNA

A in Middle of Codons		C in Middle		G in Middle		U in Middle	
Codon	Amino Acid	Codon	Amino Acid	Codon	Amino Acid	Codon	Amino Acid
AAA Lys (lysine)		ACA Thr (threonine)		AGA Arg (arginine)		AUA Ile (isoleucine)	
AAC Asn (asparagine)		ACC Thr		AGC Ser		AUC Ile	
AAG Lys		ACG Thr		AGG Arg		AUG Met (methionine)	
AAU Asn		ACU Thr		AGU Ser		AUU Ile	
CAA Gln (glutamine)		CCA Pro (proline)		CGA Arg		CUA Leu (leucine)	
CAC His (histidine)		CCC Pro		CGC Arg		CUC Leu	
CAG Gln		CCG Pro		CGG Arg		CUG Leu	
CAU His		CCU Pro		CGU Arg		CUU Leu	
GAA Glu (glutamic)		GCA Ala (alanine)		GGA Gly (glycine)		GUA Val (valine)	
GAC Asp (aspartic)		GCC Ala		GGC Gly		GUC Val	
GAG Glu		GCG Ala		GGG Gly		GUG Val	
GAU Asp		GCU Ala		GGU Gly		GUU Val	
UAA Stop		UCA Ser (serine)		UGA Stop		UUA Leu	
UAC Tyr (tyrosine)		UCC Ser		UGC Cys (cysteine)		UUC Phe (phenylalanine)	
UAG Stop		UCG Ser		UGG Trp (tryptophan)		UUG Leu	
UAU Tyr		UCU Ser		UGU Cys		UUU Phe	

The central dogma of molecular biology can now be explained. The DNA code is read in segments, and transferred to a corresponding strip of mRNA. The start codon is used to tell the cell machinery exactly where to start reading the base pairs so that the codons would be read correctly. If the cell machinery started reading one or two bases away from the start codon we would have what is called a *frame shift*, and the rest of the codons would be read incorrectly. As the codons are read correctly, starting from the right *frame*, the proper amino acid is added to the growing amino acid chain to build the protein. The protein is used to do a specific cellular task or build a certain tissue that gives the living creature its size, shape form, characteristics, and inheritable traits. So the flow of information in life is genes, which contains DNA, which codes the genetic code which codes for the proteins of life, these manufactured proteins determines the living organisms characteristics, and inherited traits. So inherited traits come from the chromosome, which contains the genes, which contain the DNA, which contain the code of that creature's life, and how it varies with each successive generation.

Thus DNA was seen by the first molecular biologists as the controlling entity of life, the central metaphor that we use to explain how life works; and so became entrenched in the modern western mind. We are controlled and determined by our DNA, our ultimate master. But is this really true?

Note two things: metaphors in science, and DNA controlling our destiny. Since when did objective factual science need to use *metaphors* to express what it needs to say? Since when does the recording tape or CD rule and control the tape player or CD player? We buy CD players first, choosing the features we want, and then we stock up on the music CDs or downloaded songs that we wish to play. Since when did DNA become the most important and necessary component in the play of life? Without the machinery of life to read, decode and manufacture from DNA, the DNA strand becomes worthless. DNA without its decoding machinery is a molecule with a message that will never be used, sitting around doing nothing. Have we been misled by a very effective but essentially incorrect metaphor/story/explanation/theory? We will deal with this later, but now is a good time to return to chemical evolution, and realize the staggeringly enormous task a materialist has in explaining life as the result of random chance via material means.

6

Primordial Soup

THE LAST CHAPTER PRESENTED a quick overview of chemistry, biochemistry, and molecular biology. I hoped to give the non-specialist reader enough background to appreciate how intimidating chemical evolution can be to a materialist's explanation.

In chemistry we have two terms—reducing and oxidizing, that describe a great deal of chemistry. Most chemical reactions can be classed into these two basic groups. Oxidizing reactions are reactions that require oxygen as one agent, or in a more general sense the oxidizing agent or chemical is the one that gives up electrons to the other agent or chemical with which it is reacting. Reducing reactions are reactions where one chemical accepts electrons from the other chemical. So for a huge number of chemical reactions one chemical is reduced and accepts electrons from another chemical that is oxidized that gives up a few electrons to the other chemical. Chemical compounds that are likely to give up electrons in a chemical reaction are called oxidizing. Compounds that are likely to accept electrons in a chemical reaction are called reducing. Most of the universe is filled with hydrogen and other such compounds, and creates what we call a reducing chemical environment. The oxygen atom or diatomic oxygen molecule is very oxidizing. Oxygen gas in a chemical environment will oxidize everything with which it reacts. Chemical evolution requires a chemically reducing environment for any of the reactions to work. Introduce monoatomic or diatomic oxygen into the environment, and all of the basic chemical evolution reactions will not go. You will get other compounds; molecules that do not give you the molecules of life, and the whole chemical evolution scenario will fail. Thus discussions of atmospheric oxygen present in the early earth are crucial. If oxygen is present in sufficient quantities, the whole scheme is a flop, and we have no naturalistic explanation for the origin of life.

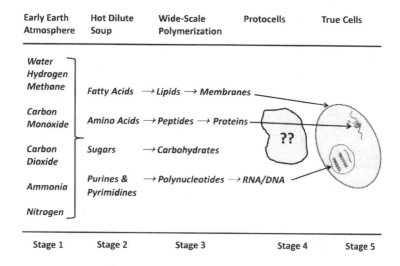

Above is diagram that has been redrawn and slightly modified from a similar diagram in Bradley, Thaxton, and Olsen's book *The Mystery of Life's Origin*[1]. It illustrates the major stages of chemical evolution. Stage 1 starts out with an early earth atmosphere that is chemically reducing, composed of water, hydrogen, methane, carbon monoxide, carbon dioxide, ammonia and nitrogen. Stage 2 is where the molecules of life are synthesized in the primordial earth's ocean to form a hot dilute soup (primordial soup). Stage 3 is where the basic molecules of life are assembled to more complex molecular structures and the parts needed for life. Stage 4 is where these parts are assembled into a simplified (if possible) self-reproducing machine or protocell, (remember our self-reproducing 3D printer). Lastly in stage 5 is where this protocell evolves by means of Darwinian evolution to become the true cell that we understand today.

Stage 4 is where we assemble the self-reproducing machine or cell, and the serious scientific literature on how this could happen is practically nonexistent, just suggestions, just-so stories, and very abbreviated computer simulations at best. The honest specialists will tell you that this so, and argue that we just don't have the capability to envision how it could happen at

1. Thaxton, The Mystery of Life's Origin, 15.

this point, i.e. we really don't know. The following quotes from well-known scientists are very revealing.

> More than 30 years of experimentation on the origin of life in the fields of chemical and molecular evolution have led to a better perception of the immensity of the problem of the origin of life on Earth rather than to its solution. At present all discussions on principal theories and experiments in the field either end in stalemate or in a confession of ignorance.[2]
>
> Klaus Dose. "The Origin of Life: More Questions than Answers." *Interdisciplinary Science Reviews* 13 (1988) 348

> Many investigators feel uneasy about stating in public that the origin of life is a mystery, even though behind closed doors they freely admit that they are baffled. There seems to be two reasons for their unease. Firstly, they feel it opens the door to religious fundamentalists and their god-of-the-gaps pseudo-explanations. Secondly, they worry that a frank admission of ignorance will undermine funding, especially for the search for life in space.[3]
>
> Paul Davies, *The Origin of Life*, London: Penguin Books (2003) xxiv

At this point the conversation will often take a philosophical turn and arguments will be made regarding how science operates, and how successful it has been in the past. The discussion often turns rhetorical; because nobody can provide a detailed explanation as to how these structures could be turned into a self-reproducing machine/cell by any means that stands up to serious scrutiny. This is a crucial point. It turns the materialist into a very religious person. They believe that science as practiced under methodological naturalism will someday find a naturalistic answer to this deep problem. The forces and particles will somehow turn some accidental molecules of life into a self-reproducing machine that evolves into a cell. This is a presupposition of faith, a tenant of the faithful; in spite of the overwhelming evidence that argues against it, against the possibility of a primordial ocean and the molecules of life in the first place. What are some of these objections and countering evidence?

First let us look at how much time and effort has been thrown into the chemical evolution task. Charles Darwin suggested the possibility in a letter to Joseph Hooker in 1871.

---

2. Dose, "The Origin of Life", 348.
3. Davies, The Origin of Life, xxiv.

My dear Hooker,

It is often said that all the conditions for the first production of a living organism are now present, which could ever have been present.— But if (& oh what a big if) we could conceive in some warm little pond with all sorts of ammonia & phosphoric salts,—light, heat, electricity &c present, that a protein compound was chemically formed, ready to undergo still more complex changes, at the present day such matter w be instantly devoured, or absorbed, which would not have been the case before living creatures were formed.—

Yours affec | C. Darwin[4]

The first serious chemical evolution theory was proposed in the 1924 by the Russian biochemist Alexander I. Oparin. The British geneticist J. B. S. Haldane put forth essentially the same theory in the 1930s. Since it was before Oparin's work was translated and published in the West, the theory is called the Oparin-Haldane theory of chemical evolution. An interesting piece of trivia is that both were Marxists, one from Russia by default (or choice?), and the other from Britain and a Marxist by choice. Another interesting piece of trivia is that Haldane is often quoted as the primary model in the composite character of Professor Weston. The Oxford don and Christian apologist C. S. Lewis created Weston in his famous "Space Trilogy" series of science fiction books in the early 1950s. There Weston is portrayed as the materialistic scientist who "sees" mankind's highest destiny as populating the universe, and nothing more. Indeed it is very difficult to find serious proponents of chemical evolution, who do not take an atheistic or material view of life even to this day. Many people are surprised to find out that Charles Darwin held many religious viewpoints through his life. But the primary trajectory of his belief can be described as a Victorian Christianized Anglicanism that slowly gives way to atheism and agnosticism as he reaches the end of his life. Thus by picking the appropriate stage of Charles Darwin's life, you can get a variety of religious quotes from Darwin, either for, neutral or against Christianity, and non-material viewpoints. It was in the latter stages of Darwin's life that he wrote his famous letter to Joseph Hooker.

Since the thirties the field languished until 1953, when the University of Chicago Chemist and Nobel Laureate Harold Urey and his graduate student Stanley Miller tested the Oparin-Haldane hypothesis. In their famous Miller-Urey experiment, a mixture of water, methane, ammonia, and hydrogen were added to a sterile heated flask that allowed the heated

4. Darwin, "To J. D. Hooker."

vapors to pass through an electrical discharge before being condensed, and returned to the original flask. The figure below shows a diagram of the original experiment.

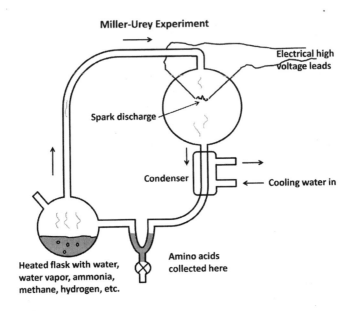

Miller-Urey Experiment

Electrical high voltage leads

Spark discharge

Condenser

Cooling water in

Heated flask with water, water vapor, ammonia, methane, hydrogen, etc.

Amino acids collected here

In this experiment, the gases reacting under the electrical spark produced a red resinous mixture that was initially analyzed to show the presence of five of the essential amino acids, and latter experiments were analyzed using modern equipment to show trace amounts of all 20 essential amino acids and even more of the non-protein building amino acids[5]. This was a stunning result in 1953 and greatly bolstered the hopes for a materialistic origin of life. Images of the Miller-Urey experiment have been gracing the pages of biology and biochemistry textbooks for decades and it is widely used to instruct science students in how "life" began. Simple sugars like the ribose in DNA and the heterocyclic base pairs of DNA were subsequently synthesized in similar origin of life experiments. However what is *not widely discussed* is what has happened since the early 1953 success. Miller and other researchers tried to move on to stage 3 of chemical evolution, the polymerization (or hooking together) of these simple molecules to form the long biopolymers necessary for life. It is here that chemical evolution

5. Parker, "Primordial synthesis," 5526–31.

has stalled and stalled badly. Making the step to stage 4 the first protocell is so beyond our understanding that it is rarely discussed except in the most general non-scientific terms. Now is a good time to discuss the objections to chemical evolution as they stand today. Remember the sequence discussed in *The Mysteries of Life's Origins*:

Stage 1—Early Earth Atmosphere

Stage 2—Hot Dilute Soup

Stage 3—Wide-scale Polymerization

Stage 4—Protocells

Stage 5—True Cells

We will dwell mainly on stage one, two, and three of chemical evolution as there is little to discuss in stage four and five. These objections come from a variety of sources but mainly from *The Mystery of Life's Origins*. Some are easy to follow, and others very technical, the non-specialist can just skip over the technical points, and still get the impact.

1. *Life appears in the geological record almost as soon as the earth had conditions possible for its survival.* Thus the vast amounts of time necessary for the production of the incredible complexity of the first cells is not there. The problem is so severe that it led Francis Crick, Nobel Laureate, and co-discoverer of the structure of DNA, to "seriously" propose Panspermia, (the idea that life was seeded on the earth by extraterrestrials, or from elsewhere in the universe) in a vain attempt to escape the time problem. This was in the early seventies where it was thought that billions of years were available for chemical evolution. Recent evidence[6] has pushed the presence of life back to 3.77 to 4.28 billion years ago. Actual micro-fossils *very* similar to modern fossils have been dated back as far as 3.5 billion years ago. Hugh Ross in his book *Improbable Planet*[7] discusses this in detail, and introduces evidence for the late heavy bombardment (LHB)[8,9,10]. This was a period in the early earth's formation where an inward movement of Jupiter in its orbit around the sun caused a vast array of asteroids inside the orbit of Jupiter to be perturbed so that they pummeled the earth, moon, and

6. Dodd, "Evidence for early life," 60.

7. Ross, Improbable Planet.

8. Tsiganis, "Origin of the orbital architecture," 459–61.

9. Morbidelli, "Chaotic capture," 462–65.

10. Gomes, "Origin of the cataclysmic," 466–69.

other inner planets. This enormous bombardment of asteroids and meteors is thought to have given rise to many of the crater features we see on the moon (the Apollo landing site craters have similar dates of formation). This "Late Heavy Bombardment (LHB)" is thought to have delivered *tons of material per square meter* of the earth's surface during the period of 3.92 to 3.85 billion years ago. While not proven there is strong evidence for this period. No life could have survived in such a torrential environment. This gives us an incredibly narrow window for the first cells to form from the theoretical pre-biotic soup; ~ 0 to 200 million years. If one dates the dawn of life to known micro-fossil dates, it "might" be pushed as high as 500 million years. Either way you cut it, the time available to randomly generate even a simple protein is unbelievably short. One can seriously argue that as soon as life was possible on earth it was there.

2. *No geologic record of the pre-biotic soup.* There is no deep layer that is filled with the mineralized remains of the pre-biotic soup. Some will argue plate tectonics and time have erased all vestiges of this layer. But in layers dated to this early era that are accessible to geologic study there are only hints of a pre-life fossilized ocean of biological chemicals. Those hints imply the heavy impact of the Concerto effect (discussed in point 8), not the chemicals one would expect to find in such a layer. The quote of Brooks and Shaw in *Origin and Development of Living Systems* as quoted in *The Mysteries of Life's Origins* is revealing.

> If there ever was a primitive soup, then we would expect to find at least somewhere on this planet either massive sediments containing enormous amounts of the various nitrogenous organic compounds, amino acids, purines, pyrimidines, and the like, or alternatively in much-metamorphosed sediments we should find vast amounts of nitrogenous cokes [graphic-like nitrogen-containing materials]. In fact *no such materials have been found anywhere on earth.*[11] (Emphasis added)

3. *No data to support that a high enough concentration of chemicals was possible* in the pre-biotic soup[12]. Not in the geologic record, and all known possible mechanisms provide only very dilute concentrations of bio-chemicals.

4. *UV—ozone dilemma.* Life needs protection from the harsh ultraviolet radiation of the sun that would destroy it. UV light is commonly used

11. Brooks, Origin and Development of Living Systems, 5366.
12. Thaxton, Mystery of Life's Origins, 58, 60.

to sterilize in the medical industry. Early life would be destroyed by direct exposure to sunlight containing this UV light. Today the O_3 or ozone layer in our upper atmosphere filters out this very harmful ultraviolet radiation. But to get ozone in the upper atmosphere you must have O_2 or oxygen in the atmosphere. This oxygen molecule in the upper atmosphere is broken apart by the harsh ultraviolet light, and recombined with another O_2 molecule to form O_3 or ozone. This is UV—ozone dilemma, to keep from destroying early life you need ozone to filter out the UV light, but to get ozone you must have oxygen present in the early pre-biotic atmosphere. But if oxygen is present, then the early earth's atmosphere is not chemically reducing, and the whole chemical evolution scenario will not work. It is a catch 22, a dilemma; no ozone, no life, but to get ozone you must have oxygen present, hence no life[13]. Some have argued for pockets and places where the sun's unfiltered ultraviolet light cannot penetrate; like the mid-oceanic hydrothermal vents or dark caverns. But this reduces the number possible sites for chemical evolution to take place to orders, and orders of magnitude smaller sites.

5. *Inability to form protein peptide bonds in the presence of water without a cellular mechanism.* In general chemistry we teach Le Chatlier's principle. This says that a reaction in equilibrium will react in the opposite way so as to relieve the stress or pressure put on it. Take a simple reaction:

$$A \longleftrightarrow B$$

If we add more A to the reaction mixture the reaction will go to the right to relieve the stress and reach a new equilibrium. If we add more B to the reaction mixture the reaction will go to the left to reach a new equilibrium point. Now take two amino acids (or more) that we are bonding together to form a peptide:

$$\text{Amino Acid 1} + \text{Amino Acid 2} \longleftrightarrow \text{Peptide} + H_2O$$

Every time two amino acids bond together an H from one amino acid and an OH from the other amino acid is removed and released as a water molecule when the dipeptide is formed. In chemistry we call this a dehydration-condensation reaction. Add another amino acid to that dipeptide chain and another water molecule is formed. If I add a whole lot of water to the reaction mixture which way is the reaction going to go to relive the stress? The answer is obvious, to the left side of the

13. Thaxton, Mystery of Life's Origins, 87.

reaction equation where no peptide is formed. This is not a trivial point. Inside the cell carefully constructed protein structures exclude or manage the excess water so that the peptide forming reaction can proceed to the right. But in the pre-biotic ocean no such cellular mechanism exists, just a whole lot of water to ensure that peptides and proteins just don't naturally occur in the "pre-biotic" soup. In an unusual experiment at Lawrence Livermore National Laboratory[14] a metal pellet containing amino acids in water solution were encased in a metal casing. An impactor was shot through a special high speed space gun that shot the impactor into the metal casing of amino acids at ~4,600 mph. This produced incredible pressure on the solution inside. On retrieving the metal casing remains and examining the solution inside they found some of the amino acids bonded together due to the incredible shock wave pressures produced on the solution. Jennifer Blank, the leading scientist, was able to produce a few dimers (two amino acids bonded together) and a few trimers (three amino acids bonded together). But really; are we are reduced to hoping that meteorite impacts into the early earth at stunning "burnt to a crisp" velocities is going to create all our needed proteins? This is how little progress has been made in even connecting amino acids together, i.e. stage 3 of chemical evolution.

6. *Inability to form or select non-racemic or optical pure compounds necessary for biological construction.* Some organic molecules have this unique ability to rotate plane polarized light clockwise or counter-clockwise a few degrees if they contain what is called a chiral center. A chiral center is when a central atom (usually carbon) is bonded to four different groups. What is unique is that there are two different ways you can arrange these four groups around the carbon. These two different ways cannot be superimposed on each other. It is like your right hand and your left hand. Try to rotate your left hand so that it exactly matches up and can be super-imposed on your right hand, you cannot do it. One of these ways (or *stereoisomer* as we call it in chemistry) rotates polarized light to the left or counter-clockwise, the other *stereoisomer* rotates polarized light to the right or clockwise. In the organic synthesis chemistry lab we almost always get a random equal mixture of both stereoisomers or a *racemic* mixture. State of the art organic chemistry is working with chemical reactions and procedures that will get you "mostly" (80 percent is a great result) one *stereoisomer* in a few very select reactions. Life requires 100% stereoisomer pure reactions and compounds that all rotate light the same way: *levorotatory* or the

14. Blank, "Hitchhiker's Guide to the Early Earth".

l *stereoisomer*[15]. We cannot do this in the lab, but life and all of its compounds do this at 100 percent purity all the time. How did such a stereo-specific process come to be? No one has any hard evidence or any answer that isn't pure speculation.

7. *The lack of discussion of interfering reactions that must be present in the pre-biotic soup* that would surely tie up any significant production of biologically useful compounds[16]. The Oparin-Haldane theory ignores all reactions that can occur in the reducing environment that will not give you any useful biological compound, but in fact tie up the production of useful biological compounds.

8. *The Concerto effect, the energy that drives creative reactions is just as likely to drive destructive reactions*[16]. It takes an input of energy to get these chemical evolution reactions going; either in the form of heat, light energy, shock (intense pressure) waves or electrical energy. But the energy that drives the production of amino acids, etc., will just as likely drive reactions that tear these compounds apart or make then react to something else that is not biologically useful.

9. *Lack of thermodynamic support for the massive configurational energy required in the thermodynamically open systems* demanded by all chemical evolution theories to account for the enormous complexity of the cell. Classical thermodynamics does not favor the reaction scenarios needed to go from soup to cell. The soup to life process flies in the face of simple and complex thermodynamics. We could summarize it as "a tornado going through a junkyard does not produce a Boeing 747 jet." The knowledgeable materialist concedes this is a real problem yet to be solved. The non-thermodynamically trained biochemist or biologist usually misses the point. Of all of the natural sciences, thermodynamics is considered the most reliable and stable. To quote the famous astronomy professor Sir Arthur Eddington, who verified one of Einstein's theory of relativity predictions:

> The law that entropy always increases, holds, I think, the supreme position among the laws of Nature. If someone points out to you that your pet theory of the universe is in disagreement with Maxwell's equations—then so much the worse for Maxwell's equations. If it is found to be contradicted by observation—well, these experimentalists do bungle things sometimes. But if your theory is found to be against the second law of thermodynamics

15. Thaxton, Mystery of Life's Origins, 52.
16. Thaxton, Mystery of Life's Origins, 47.

I can give you no hope; there is nothing for it but to collapse in deepest humiliation.[17]

Open thermodynamic systems, like the early earth is postulated to be, provide the possibility of going against the degenerative direction of the second law of thermodynamics; but the degree expected by chemical evolution far, far exceeds anything ever seen in the physical world, and by what can be justified by quantitative calculations[18].

10. *Problem of the potential massive irreducible complexity of most biochemical systems.* Dr. Michael Behe publicized this concept in his famous book, *Darwin's Black Box*. It will need a separate section of its own to explain it.

11. *The likely presence of oxygen in the early atmosphere that would completely stop all biochemical relevant reactions in the soup.* Modern geologic evidence is tipping towards the presence of some oxygen in the early earth's atmosphere[19]. If water is present, which we know it surely will be, then in the upper atmosphere the ultraviolet light from the sun will separate water, H_2O, into its constituent hydrogen, H_2, and oxygen, O_2. Thus there has to be some oxygen present in the atmosphere. How is it excluded to prevent it from stopping the reducing chemistry needed by the Oparin-Haldane theory? Recent evidence strongly implies that oxygen at fairly significant levels has been present since the very earliest times of the earth's formation[20]. Or as Bruce Watson, Institute Professor of Science at Rensselaer says:

We can now say with some certainty that many scientists studying the origins of life on Earth simply picked the wrong atmosphere[21].

12. *The inability to separate investigator interference from most origin of life experiments*[22]. A quick look at the earlier mentioned Miller-Urey experiment reveals a heater, leak-tight container, tubes that conduct the heated vapors pass a carefully located electrical spark, and then on down to a collection pool where sample are collected separate from the initial heater and mixture of heated chemicals. This arrangement of coordinated parts bears the marks of an intelligently

17. Eddington, The Nature of the Physical World, 74.

18. Thaxton, Mystery of Life's Origin, chapter 7 and 8.

19. Thaxton, Mystery of Life's Origin, chapter 5.

20. Trail, "The oxidation state of Hadean magmas", 79–82.

21. Rensselaer Polytechnic Institute. "Earth's Early Atmosphere".

22. Thaxton, Mystery of Life's Origin, chapter 6.

designed device engineered by the investigator. Where in the pre-biotic ocean do we find such coordinated arrangements of orifices and containers to carry out such an experiment? Who watches the time, the heat level, the spark duration, and sampling frequency in the natural pre-biotic ocean? How well does this experiment really match the conditions one would really find in the pre-life pre-biotic ocean? Many have suggested—not well at all.

13. *Eigen's paradox and the error threshold*[23]. Once the pre-biotic soup has "theoretically" progressed to the level of self-replicating molecules a serious problem arises. RNA has been proposed as the likely candidate molecule since it known to reproduce itself partially under controlled experiments. However to "mutate and evolve" a molecule must make mistakes in copying itself. But to past on beneficial mutations to the next generation of molecules the information contained in that strand of RNA must be copied faithfully. If not, all of the information including the beneficial mutation is degraded and lost. Thus for a self-replicating molecule there is a critical balance between the error rate in copying, giving the possibility of mutation, and total information copying fidelity, that preserves it and the other information content present. To get the copying fidelity needed in DNA and RNA replicators to meet this balance, the replicating mechanism must have some sort of error correcting mechanism. This requires a very, very large DNA, or RNA molecule, or protein, and is very much larger that a RNA strand composed of only 100–300 RNA units bonded together. In the seventies Manfred Eigen and Peter Schuster showed that without some sort of error correcting mechanism you cannot get a self-replicating RNA molecule larger than 100–300 RNA units long before the copying errors will hinder the replication sufficiently to degrade the information in the molecule. But to get past this 100 to 300 RNA bases (units) limit you need an error correcting mechanism which requires a RNA strand very, very much longer. Thus we have a chicken and egg problem. Which came first: the very large RNA or DNA strand (and proteins??) with error correcting functions; or the very much smaller RNA strand you must evolve to get to the possible error correcting very much larger RNA strand. How do you bridge this gap? No one has any good ideas.

14. *The total lack of any serious evidence that information content found in DNA can be produced apart from an intelligence that produces it.* This is extensively discussed in Stephen Meyer's book, *Signature in the Cell*[24],

23. Eigen, "Self-organization of matter," 465–523.
24. Meyer, Signature in the Cell.

and merits a separate section to discuss it. We have touched on this topic in our discussion of the self-replicating 3D printer. Where does the program to run it come from, including all error correction protocol? How do you get a program of this sort from a non-intelligent source?

Point 14 was a critical point that persuaded the well-known agnostic intellectual Anthony Flew to convert to a theistic (though non-Christian) position before he died. It is such a profound point that Stephen Meyer wrote an entire book, *Signature in the Cell,* based on it. Michael Polanyi is credited with being the first person to publicize the significance of information in the cell[25] to the origin of life debate. As mentioned earlier, he explained how to the extent a biological organism can be shown to be a machine, to that extent it cannot be reduced to the laws of physics and chemistry. He was one of the first to realize that the genetic code contained along the spine of a DNA strand was information, and information that was independent of the medium that stored it; be it a strand of DNA, a piece of paper with GCAT's all written down, or a computer memory that stored the genetic code in its memory banks. For information to be information it has to be independent of the medium that stores it. It has no physical property in and of itself, though it has the possibility of producing many physical properties when decoded and implemented.

Charles, Ellie, Mary, Kristina, and Jon were gathered in the back dusty faculty conference room as the end of semester exams loomed darkly in the near future. The studying was done and the conversation drifted as usually happens.

"Hello there? You gals have to actually study for your exams?" Bert stuck his head around the door, and quietly drifted in.

"And I suppose you don't," replied a slightly amused Mary.

"I wish," he replied. "However some girls do not seem to need to. Right?"

"Yea, right," Mary blushed slightly.

"But you do it anyway to keep the rest of us from looking bad"

"Like you should talk," but Mary was trying very hard not to show real color as she had picked up that Bert was not being sarcastic. With a flip of her hair she turned to face Bert and asked, "Maybe you can explain just what DNA information is. It is not like it is a book that you read or something."

"Well I can try," Bert turned to the nearest dry erase board, grabbed a green marker, and drew the following:

25. Polanyi, "Life's Irreducible Structure."

"Okay what is that?" cried Kristina.

Bert eyed both Mary and Kristina and started explaining. "Each string of five boxes is a stretch of DNA with let's say five codons where three base pairs specify a unique codon. So I am really looking at a stretch of 15 base pairs along a strand of DNA. But to generalize things a bit let's just say we have 26 letters of the alphabet instead 20 essential amino acids to code for, and stop and start codons. Each of these groups of five letters has the possibility of carrying a certain amount of information, not an encyclopedia's worth, but a useful bit. This is called Shannon information after the information scientist Claude Shannon who first identified it." Bert went over to the board and wrote some more words to the side of the boxes.

"Now I have labelled each box as to what kind of information it might likely contain. I use Shannon information on the first, third, fourth, and fifth box, because though they might contain useful information, according to the rules, grammar and words used in the English language they appear to be a random set of five letters with no meaning attached to them. However the second box has the English word hello written in it with all the meaning and purpose that hello conveys. Thus we call the second word or group of five letters *specified* information. Here I am using the word Shannon information in the sense that Steven Meyer did in his book *Signature in the Cell*, an information carrying capacity.

| A | D | E | G | T | Shannon information |

| H | E | L | L | O | specified information |

| B | S | L | I | O | Shannon information |

| A | A | T | G | B | Shannon information |

| L | L | M | O | E | Shannon information |

But the concept of specified information is generally understood in the information community as a grouping that conveys meaning; that directs; commands, etc."

At that point Bert stepped over to the board, erased the Shannon and specified information words, and then marked in some fractions and equations as follows:

| A | D | E | G | T |

$$\frac{1}{26} \cdot \frac{1}{26} \cdot \frac{1}{26} \cdot \frac{1}{26} \cdot \frac{1}{26} = \left(\frac{1}{26}\right)^5 = 0.0000000842$$

| H | E | L | L | O |

$$\frac{1}{26} \cdot \frac{1}{26} \cdot \frac{1}{26} \cdot \frac{1}{26} \cdot \frac{1}{26} = \left(\frac{1}{26}\right)^5 = 0.0000000842$$

| B | S | L | I | O |

$$\frac{1}{26} \cdot \frac{1}{26} \cdot \frac{1}{26} \cdot \frac{1}{26} \cdot \frac{1}{26} = \left(\frac{1}{26}\right)^5 = 0.0000000842$$

| A | A | T | G | B |

$$\frac{1}{26} \cdot \frac{1}{26} \cdot \frac{1}{26} \cdot \frac{1}{26} \cdot \frac{1}{26} = \left(\frac{1}{26}\right)^5 = 0.0000000842$$

| L | L | M | O | E |

$$\frac{1}{26} \cdot \frac{1}{26} \cdot \frac{1}{26} \cdot \frac{1}{26} \cdot \frac{1}{26} = \left(\frac{1}{26}\right)^5 = 0.0000000842$$

"Before we go any further, I want to make a point."

"Okay," replied Mary.

"If you look at the probability of each word occurring by chance you notice that even for a short five letter word the probability is getting small, and it grows exponentially with the size of the word. In others words we are seeing combinatorial inflation again. But the crucial point is that every five letter word has the same small probability whether it is a meaningful word or just gibberish. And this will still apply when we expand the word to a hundred letters with astronomical small probabilities. The gibberish 100 letter word is just as unlikely as the meaningful word."

Katrina eyed Bert suspiciously, "What does this mean then?"

"Michael Polanyi in his book *Personal Knowledge*, pointed out that the remarkable aspect is not the small probabilities involved, but the fact that we recognize the word HELLO, and understand what it means. To recognize a word implies a language with its grammar, rules, forms, words, and the culture, minds and civilization that created it all. There is so much wrapped up in the act of recognizing a word and understanding its meaning that it boggles the mind. We don't even understand the process by which the image recorded by the eye and sent to the brain is decoded to the greeting, Hello!"

Ellie jumped in, "It's a form of tacit knowledge, isn't it?"

"Yes, I think you are right, specific information when it is seen implies a lot, an intelligence of some sort. Generally we never see information in the natural world unless it originates from an intelligence or mind."

"Really?" said Jon.

"Really!" said Bert. "Steven Meyer in *Signature in the Cell* phrased it this way."

> Apart from the molecules compromising the gene-expression system and machinery of the cell, sequences or structures exhibiting such complexity or specified information are not found anywhere in the natural—that is, the nonhuman—world.[26]

"Do you really memorize long quotes like this?" Ellie was looking at Bert with a certain degree of amazement. "Do you have a photographic memory?"

"No I don't, but I can picture short sections of text for a few days after I have read it, and then it fades away, sometimes."

"But just what is a word in the human genome or genetic code," asked Mary?

"That is a very good question," replied Bert. "We know that the start and stop codons are commands, "words," but beyond that we know that a particular protein does a certain function. So is the string of codons or amino acids that codes for that protein a "word" command or what? Actually, many specialists tend to think of the genetic code and genome as a program, an algorithm, with an operating system of some sort; and the database of needed proteins and whatever else is expressed by the genome." That is why the bioinformatics specialist Hubert Yockey said:

> The genetic code is constructed to confront and solve the problems of communication and recording by the same principles found . . . in modern communication and computer codes.[27]

"You are showing off again, did you really memorize that?" Ellie was pestering Bert with "Really?" type of look.

"No, Ellie I read them from the side wall where I post my study notes."

Ellie glanced at the wall then quickly looked back as she realized she had been caught by a grinning Bert, Charlie, and Jon. "Whatever," she said.

"Continue on Bert," Katrina glanced at the guys and then at Mary who seemed to be losing interest.

"Well it gets even more interesting. Apparently for many proteins the codons that code for it are not always contiguous. There are sections of

---

26. Meyer, Signature in the Cell, 110.

27. Yockey, "Origin of Life on Earth," 105–23.

non-coding codons interspersed between the sections of coding codons or base pairs. The sections or stretches of DNA that code for the protein are called *exons*, (for "expressed" regions), and the sections that the reading machinery has to skip over to find the next exon are called *introns*, (for "interspersed" regions)."

"Well how much of the human genome is exons, and how much is introns? Ellie was fascinated at the complexity emerging from Bert's description. Jon was looking stunned; he almost looked a bit pale.

"The National Institutes of Health funded ENCODE project seeks to find out how much bio-function exists in the human genome, i.e. the human DNA. After years of work they homed on ~1.5 percent of the human genome directly codes for proteins[28]. The rest does nothing or something else. Fifteen years ago the other 98.5 percent of the genome was considered junk DNA. Pieces of leftover virus's DNA inserted generations ago, evolutionary dead-ends, and flotsam. The sort of pattern you would expect if the genetic code had to evolve over eons with many dead-ends, trials, and failures. What they have found since the earlier days is stunning."

"What is that?" asked Jon who was leaning forward intently.

"Let me try to reproduce a diagram from the 2014 paper[29] they published," Bert strode over to the board, erased the previous diagram, and drew a series of circles of different colors. "The outside black line represents the entire human genome. The little dark gray circle represents the exons that we know code for specific proteins. In 2000 the rest was called "junk" DNA because it could not be correlated to any protein and was full of short snippets and a lot of repeating sequences. Thus by "definition" it was not useful, and arose from our evolutionary past. Thus was the power of the Central Dogma metaphor. From DNA genes to proteins to you—that was the story. Scientists quickly figured out that this was way, way too simplistic and maybe just flat wrong. Suspicions arose about the rest, the ENCODE project was born, and now we have this.

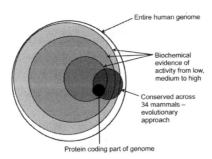

28. Kellisa, "Defining functional DNA," 6134.
29. Kellisa, "Defining functional DNA," 6132.

The light to mid gray dotted circle regions which are pushing over 80 percent of the genome have showed signs of biochemical activity ranging for low to high. The mid gray solid circle region includes most of the protein-coding regions and other parts of the genome that appear to be "conserved" (stays the same) across 34 different mammal species. Computational biologists call this the evolutionary approach where you try to establish evolutionary linkages by percent similarity of certain proteins and DNA sequences across the different but evolutionarily related species. I have often wondered if you could have done the same thing by assuming a functional design approach, i.e. similar design requires similar functionality in the mid gray solid circle area. The intelligent design concept implies this strongly. But the evolutionary metaphor is just too firmly ingrained for those with the resources to try another approach in the mid gray solid area.

The dark gray smallest solid circle region is what most people think of when they talk about the genomic code; it is the part that actually codes for proteins. But one thing is becoming apparent, there is very little if any junk DNA in the human genome." If it is there it will likely be in the white region of the entire human genome circle. If fact, the head of the ENCODE project Ewan Birney said in an interview with the magazine Scientific American; whoa, wait! Here is the actual copy of the magazine here." At that point Bert walked across the room to a low table by a chair and picked a copy of a magazine, showed the cover to everyone and then flipped through a few pages to read:

> *Scientific American Magazine question*—Should we be retiring the phrase "junk DNA" now?
> *Ewan Birney answer—Yes, I really think this phrase does need to be totally expunged from the lexicon.* It was a slightly throwaway phrase to describe very interesting phenomena that were discovered in the 1970s. I am now convinced that it's just not a very useful way of describing what's going on.[30] (emphasis added)

Mary looked up and quickly asked, "Does everybody agree with the ENCODE project?"

"No, some scientists got very upset at them, and have published critical critiques of their methods, but the evidence for their results is getting stronger, not weaker, as we research the human genome. Like Birney says in this article; everybody "knew" that the old model was just too simplistic, and that a lot more was going on."

---

30. Hall, "Hidden Treasures in Junk DNA."

Jon was very pensive as he looked at Ellie and whispered loudly, "Ellie, can we go out somewhere and talk?"

"Sure."

Charles stood up, stretched, "I'm tired and need to go to bed!" Like a cue from a movie film, the gathering broke up and filtered out the door chatting quietly as they left.

So then, what is a "word," sentence, or program command in the genome code? We really don't know at this point. We are not even sure if those terms apply as we currently understand them. The human genome seems to be beyond our concepts of language and programming. Experienced programmers who have worked with the genome sometimes comment that it "feels" like they are trying to understand Windows OS version 20 or something[31]. Vast areas of human genome knowledge are being glossed over in this book, but there is one incredible aspect of the genome that needs discussion before we move on. It stunned me and my Philosophy of Science honors student some fifteen years ago. The actual conversation is long forgotten, but the substance occurred as follows.

I was sitting in my office grading papers as professors of my sort often do. It was my second or third year of co-teaching the honors seminar Philosophy of Science with my theology co-teacher. The first year teaching of the class had been a success but we were still working the bugs out of the unusual format we had chosen for the class. I was flushed with excitement of the possibilities of the newly developing intelligent design movement that had recently sprung on the scientific and political scene in the US. Many agreed with me, a significant number did not. She knocked on the door and asked to come in.

"Dr. Collier"

"Hello! How did your summer research at OU go this summer?" She was brilliant, a double major in Chemistry and Biology, and an honors fellow who had been one of our guinea pigs in the first class of Philosophy of Science students. Never had we had a class that was so verbal and willing to discuss things. Asking them a leading question was like throwing a bone with meat on it to a pack of starving dogs. The resulting discussion was ferocious and intense, not always well-informed or thought out, but they were bright young undergraduates, eager to prove their intellectual mettle.

---

31. email sent by a genomic researcher to a listserv in which the author participates.

"It went well, very well," she replied. She glanced around my office and then started in. "You are not going to believe what I found out this summer."

She had received an undergraduate summer research fellowship from the University of Oklahoma and had just spent the summer working there. Larger state universities frequently use summer fellowships to recruit bright undergraduate students from other universities into their graduate programs.

"Well, what was it?" I asked; my curiosity piqued at her remark.

"We were working on death genes as part of our research on cancer."

"What's a death gene?"

"It is the part of DNA that codes for a protein that tells a cell that it needs to die. Cells in our body need to periodically die off and be replaced. Some think cancer is linked in to a failure of that mechanism; a cell just keeps on reproducing and growing out of control. That is why we are studying this. There is a protein called p19ARF that tells a cell to die, and we were studying the region of the DNA that produces it."

"Okay, I follow you"

"Here is what is so amazing, if the DNA does a frame shift, and reads the same gene again, you get a different protein"

"Wait!" I asked. "Explain frame shift to me."

"That is when the DNA reading mechanism goes to the start codon and before it starts to read the next codon, the next three base pairs; it skips one base pair, and then starts to read the codon."

"Wait, a minute," I said. "That means every codon read after that shift will probably be different, you will get a completely different protein, the primary amino acid sequence will be completely different!"

"Yes, Yes! And here is the incredible part; that new protein will tell the cell to die, and do it even better that the first protein will. It basically does the same thing but better!"

I grabbed a piece of paper and with her looking over my shoulder I scribbled.

... ATG TGT GAT GCT ACC CTA TGT CCA AAA GGG CAC CTG CCA ATA ACC

P1: Ile—Cys—His—Arg—Thr—Leu—Cys—Pro—Lys—Gly—His—Leu—Ile—etc.

.. A TGT GTG ATG CTA CCC TAT GTC CAA AAG GGC ACC TGC CAA TAA CC

P2: Cys—Val—Met—Leu—Pro—Tyr—Val—Gln—Lys—Gly—Thr—Cys—Gln—etc.

"Okay the top line is the sequence of bases coded into the DNA strand. The next line is the protein amino acid sequence you would get if you start decoding the first codon ATG, second codon TGT, into the first and second amino acids of the protein, isoleucine, cytosine, etc. That protein is represented in the second line denoted by P1. However if we shift by one base pair and start with the codon TGT, and then GTG as shown in the third line where we have frame shifted the reading of the codons by one base pair to the right: we would get cytosine and valine as the first two amino acids in the new protein P2." I looked up playing the professor explaining a concept to a student who already knew it when I was really explaining it to myself. "That second protein should be junk. The vast majority of protein sequences we know are not functional, so a frame shift ought to give us junk, as functional proteins are not that common. But this new protein P2 is so completely different, not a simple variation of a few amino acid, but a complete change all the way through. This is very surprising."

"It is amazing," she said. "It is not only functional, but *it does the same job and even better!* How in the world could a system like this arise by chance? How in the world?"

It was then that the full magnitude of what she was saying struck home. How in the world? This was coding and decoding on a level that the English language cannot accomplish. Let me explain with an example. Take a simple paragraph where the words are like codons and the spaces between the words indicate the starting and stopping points for those codons. Now slide the spaces between the words over one space and try to read the "new" paragraph.

Paragraph before frame shift

Those who assume that the future will be a continuation of the present suppose that modernism will always retain its worldwide dominance, although localities may be temporarily taken over by Islam or some other fundamentalism. On the contrary, I predict that the foundations of modernism will be profoundly shaken in the twenty-first century as the public becomes aware that the actual data of science disconfirrn the ambitious claims . . .

Paragraph after frame shift

Thos ewh oassum etha tth efutur ewil lb ea c ontinuationo ft hep resents upposet hatm odernismw illa lwaysr etaini tsw orldwided ominance,a lthoughl ocalitiesm ayb et emporarilyt akeno verb yI slamo rs omeo therf undamentalism.O nt hec ontrary,I p redictt hatt hef oundation so fm odernis mwil lb ep rofoundl yshake ni nth etwenty-firs tcentur ya sth epubli cbecome sawar etha tth eactua ldat ao fs cienc edisconfir rnth eambitiou sclaim . . .

To grasp what is going on, read the first paragraph and then the second. The second paragraph makes no sense. Not a single word makes it through the frame shift with a new word that can be understood or a sentence that has any meaning. If this were the p19ARF gene, not only would the second paragraph read perfectly and make sense; but it would be the logical continuing paragraph one would expect to read next in the book!

Now granted the DNA coding mechanism is not a human language, and the analogy between words of a text and a codon is a bit strained. But it does give the reader a flavor of what is going on in this double coding in the DNA. How common is this double coding in the human genome?

> The human genome has thousands of overlapping genes. "However, the origin and evolution of overlapping genes are still unknown"[32]
>
> Human DNA contains an estimated 20,000–25,000 genes[33] in its approximately 3 billion base pairs. This is down from earlier estimates of over 100,000 genes.
>
> The Human Genome Sequence Reveals Unexpected Complexity . . . Only 1.5% of the 3.2 billion base pairs of the human genome encode protein, yet those 31,000 or so genes specify 100,000 to 200,000 distinct proteins[34]
>
> Indeed, as a result of the overlapping genetic messages and different modes of information processing, the specified information stored in DNA is now recognized to be orders of magnitude greater than was initially thought[35],[36]

The rice plant genome actually has more genes than the human genome, but the proteins expressed by the human genome are much more numerous, and it is obvious that the human genome does a lot more with each gene. It is like a different sort of coding system is being used at the next higher level. It is a level of coding that we still do not understand, and one that we cannot match in efficiency and specified complexity, even with our most advanced operating systems and algorithmic knowledge. It begs the question of how such a system came to be apart from an intelligent agency.

32. Veeramachaneni, "Mammalian Overlapping Genes," 280–6.
33. Stein, "Human Genome."
34. Lewis, Life, section 13.5.
35. Meyer, Signature in the Cell, 462.
36. Johnson, Programming of Life, 26–27.

Nothing in our scientific experience can justify such a system by chance and necessity alone.

The final comments we will make about such a system is to ask questions. What is more important: the coding system or the cellular machinery that decodes and implements the code? What comes first—the CD with the program or the computer that reads the CD; stores the program, and then runs the program? How do you have one without the other? This is a classic chicken and egg problem. Which came first—the machinery or the code? At this point no one has any serious idea on how life could have evolved in the face of such a system, and how to surmount such a chicken and egg problem.

7

Science and Metaphors

WITH HER KIND PERMISSION, I have adapted and interpreted notes and material originally presented in a lecture titled "Metaphorical Language and Science" given by Dr. Lori Kanitz, Assistant Director of the Institute for Faith and Learning at Baylor University[1]. Michael Polanyi warned of the dangers of imposing a strictly detached and supposedly objective viewpoint on all knowledge. Forcing knowledge to be expressed in detached objective terms can lead to a viewpoint that only accepts empirical knowledge as true knowledge, and that can lead to some very strange results. Polanyi warned that it leads to the discrediting of *all* knowledge, since all knowledge has at its base, many tacit components which defy empirical objectification. To better understand this, let's look at one aspect of the problem—science and metaphors.

What is a metaphor? Janet Soskice in *Metaphors and Religious Language* explains a metaphor as

> "that figure of speech whereby we speak about one thing in terms which are seen to be suggestive of another."[2]

Lakoff and Johnson, in *Metaphors We Live By*, define metaphors this way:

> understanding and experiencing one thing in terms of another[3]

1. Kanitz.
2. Soskice, Metaphor and Religious Language, 15.
3. Lakoff, Metahpors We Live By.

173

Some examples will make this clear. Here are two explicit metaphors.

Time is money.
A is B

Thy word is a lamp.
A is B

Here are some implicit metaphors.

Sound *wave*

Electron *field*

Almost all English-speaking adults understand waves as the up and down motion of water on the beachfront, and as the up and down sinusoidal motion of graph plots seen when sound or light waves are represented in books. The understanding of sound motion as a compressional sine wave would not have been possible without humans first experiencing the wave-like character of the ocean front, or ripples in a mud puddle. In fact, Neil Postman argues, "A metaphor is not an ornament. It is an organ of perception."[4] Metaphors have at least six cognitive functions:

1. Organs of perception
2. Coherent systems of perceiving
3. Vehicles for transference
4. Doors to associative networks
5. Vehicles of discovery
6. Action-guiding systems

How can a metaphor be used to perceive something, that is, how can it be an organ of perception? Let's break it into two categories and give some examples. First is what Lakoff and Johnson describe as orientational metaphors.[5]

> *Virtue is up.*[6]
> She has *high* standards.
> He is an *upright* person.
> He *sunk* to a new *low*.

4. Postman, The End of Education, 74.
5. Lakoff and Johnson, Metaphors We Live By, 14-21.
6. Lakoff and Johnson, Metaphors We Live By, 16.

They are a scum sucking *bottom* feeder.

Another category Lakoff and Johnson identify is ontological metaphors where events, activities or emotions are referred to as things.[7] A familiar example is the metaphor of a brain being a machine.[8] It gives rise to metaphors like:

> My brain is a little *rusty* today.[9]
> My thinking *ground to a halt.*
> I could see the *wheels turning.*
> They've slipped a *gear.*

Another example of an ontological metaphor is time is a container:[10]

> I can't *fit* anything more into the day.
> Will you try to *squeeze* me into your schedule?
> The day *contained* a number of surprises.

In addition to functioning as an organ of perception, metaphors create coherent systems of perception. Take the metaphor, God is our Father. It can shape and shade how we read the Scriptures. How we view our life experiences will be shaped like-wise. If our image of Father is a loving, kind, protective father, then our view of God and our approach to him will be heavily influenced in that direction. If we view father as judgmental, harsh and controlling, then it will be very hard to view God in a different light. If God is our Father is our prevailing metaphor, our very identity could be shaped by that one dominating metaphor. This touches on problems inherent with metaphors that will be discussed later.

Furthermore, metaphors allow us to transfer knowledge from domains we know well to those we know less well. Metaphors have qualities that are not immediately associated with the concept to which they are applied. Thus, we can transfer those qualities of the metaphor to the concept and perhaps better understand and extrapolate from the concept illuminated by the metaphor. We transfer qualities from the known to the unknown. The following metaphors illustrate transference:

> Neuron *firing*
> Synaptic *terminal*

---

7. Lakoff and Johnson, Metaphors We Live By, 25–32.
8. Lakoff and Johnson, Metaphors We Live By, 27.
9. Adapted from Lakoff and Johnson, Metaphors We Live By, 27.
10. Lakoff and Johnson, Metaphors We Live By, 29.

Electrical *trigger*

Trigger and firing come from the use of a gun. You *fire* the gun by pulling the *trigger* and almost instantaneously the bullet cartridge explodes sending the bullet down the barrel of the gun towards its target downrange quicker than the eye can see. When a neuron in the brain *fires* it releases a small electrical discharge from one of the connections to another, sending the electrical impulse involved in a thought rapidly down the appropriate neural network. The release of the electrical charge is very rapid like a gun *firing*. It is *triggered* like a gun by some mechanism, and the relationship of one neuron to its other neighboring neurons is similar to a busy train *terminal* where railroad lines from many cities meet together in one massive hub or *terminal*. From this terminal you can go to any destination you wish, if the connecting railroad exists. Thus, the qualities we associate with guns, shooting firearms, and railroad terminals create metaphors that attempt to describe what is happening in the brain when a neuron transmits an electrical charge from one neuron to the next. We transfer the properties of the known, shooting firearms, and train terminals to the unknown; neurons working in the brains, and learn something new about how neurons and the brain work. But is the metaphor totally successful? Can we transfer everything? Obviously not. Railroads do not exist in our brain. Passengers do not ride our neurons. Bullets are not speeding down our neural pathways. So, not all aspects of the metaphor can be automatically transferred. Though a neuron is in some respects "like" a gun firing, and "like" a railroad terminal, it is "not" a gun or a railroad terminal. The better the metaphor, the better the transference is to understand new things about the concept or item. Virginia Stem Owens expresses this as the "aptness" of a metaphor.

> If the metaphor is apt, miraculously the qualities thus transferred not only stick, but tell us something fresh about their new host. Except for direct sensory experience, the apprehension of likeness is our primary way of learning about the world[11].

Next, metaphors can be powerful tools of discovery because they create associative networks. The discovery of the modern structure of the atom provides a good venue for exploring this facet of metaphorical language.

At the turn of the century, after the discovery of the electron, the model of the atom likened the thin cloud of positive charge with tiny electrons floating it to plum pudding with raisins. The raisins were the electrons, and the pudding was the positive charge. A beam of alpha particles or sub-atomic particles shot at it would likely go right through, and maybe

---

11. Owens, "Telling the Truth in Lies", 9–12.

be slightly deflected by the "cloud" or a close encounter with an electron. Ernest Rutherford designed just such an experiment to test this hypothesis. One day Rutherford wanted to recheck some results of his assistant who was conducting the experiment. Unlike the assistant who swung the detector in line with the alpha particle source before he turned the apparatus on, Rutherford turned the apparatus on first. To his complete astonishment he discovered that the "atoms" were deflecting the alpha particles at a right angle or 90 degrees. Experimenting further he found that he could swing the detector to right beside the alpha particle source, and the test atoms would bounce the alpha particles right back at the source and detector. Rutherford explained his initial reaction as, "If I fired a cannonball at a piece of tissue paper and it bounced back and hit me." The only explanation was that all the positive charge was concentrated in a small tight nucleus that could actually repel the alpha particles in direct collisions. But now the radius of the atom would be too small, unless one could move the electrons out from the nucleus with electrons *orbiting* the nucleus like the planets *orbit* the sun. This new metaphor gave rise to further questions about how properties of the atom might be like the solar system. In other words, it created an associative network of questions and possibilities. As it turns out, electrostatic attraction functions like gravitational attraction to stabilize the orbits of the electrons around the nucleus. Orbits with larger radius from the nucleus had more energy associated with them.

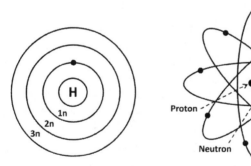

Bohr Model of the Atom   **Bohr-Sommerfeld Model of the Atom**

This Bohr model of the atom then became a vehicle of discovery. Initially all electron orbits were thought to be in the same plane, like the solar system. But later, the Bohr-Sommerfeld model reflected new understanding of the electron orbits. It depicted the orbits as occupying different planes of rotation and being more elliptical in their orbits around the nucleus.

Eventually modern quantum mechanical theory was developed and applied to the hydrogen atom. The concept of an electron orbiting a nucleus was discarded and replaced by an electron probability distribution that assumes a particular cloud-like shape around the nucleus. But these probability distributions are still called *orbitals* in deference to the initial solar system metaphor that started the whole chain of discovery.

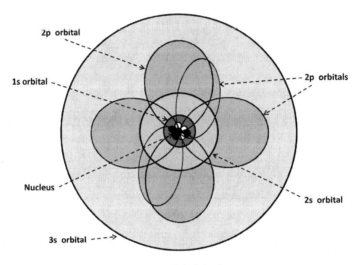

Modern Model of the Atom

Lastly, metaphors can be incredibly powerful in science and in life because they become action-guiding systems. In other words, the metaphor we unconsciously assume or consciously adopt as our model gives rise to certain actions. Lakoff and Johnson illustrate this with the metaphor *an argument is a war*.[12] If we assume that when we're in an argument, we're in a battle, the metaphor will affect all aspects of our understanding, experience, and action. The person we argue with will be understood to be an opponent. Anything an opponent in a battle does or says is an attack. And in a battle, when attacked, the appropriate action is to fight back. Furthermore, the metaphor establishes our goal—to win. We can see the evidence of this in the sorts of things we say and do when guided by this metaphor. For example:[13]

> She *demolished* my argument. (experience)
> Her criticisms *hit the mark*. (understand)
> I'll have to *try another front*. (act)

---

12. Lakoff and Johnson, Metaphors We Live By, 4.
13. Examples adapted from Lakoff and Johnson, Metaphors We Live By, 4.

However, this isn't the only metaphor we might use for an argument. What if instead we compared it to a journey? In that case, the person we argue with is not an enemy but a traveling partner. What he or she does and says is no longer perceived as attacks but guidance—perhaps critical questions and information—helpful to our journey, and our response is therefore no longer to fight back but to listen and to weigh it. Finally, our entire goal has changed. The goal is no longer to win a battle but to arrive, preferably together, at a mutually beneficial destination.

Because the metaphors we adopt affect our perceptions, interpretations of phenomena, behaviors, and goals, it is crucial in science as well as in life to examine the "metaphors we live by."[14]

A scientific example illustrating how important this is comes from studies of light at the turn of the last century. Prior to 1900, the prevailing metaphor described light as a wave. But the metaphor had several interesting properties. Waves are continuous, and you cannot locate them to any particular point in space, only along a certain directional line. The German scientist Max Planck was trying to find a mathematical formula that would predict the intensity versus frequency (color) of light emitted for a given high temperature to which a solid body was heated. This emitted light was called blackbody radiation and is what we see when we look at a tungsten filament light bulb that is turned on. He could not get the mathematical formula to fit the experimental data using the classical or light is a wave model. It was only when he introduced a constant (now called Plank's constant), that in effect quantized the energy of light into discrete bundles of energy, that he could get the equation to fit the data. The introduction of the constant was controversial in its day but proved to be very prophetic, and far-reaching in its consequences. Plank had introduced a new metaphor; light is a particle with a discrete quantity of energy. Today we call this a photon; we can calculate its energy from its frequency, and we can locate a photon to a particular point in space, even if it is moving at the speed of light. His competing metaphor started a chain of scientific discoveries that led to the modern principle of complementarity. It turns out that both metaphors are partially right, and reality is something else, but is still best described as a blending of both metaphors. The particle-like behavior or wave-like behavior of light will manifest depending on what experiment is being run. Shine white light through a prism and it separates, according its *wavelengths* into the colors of the rainbow like the wave model predicts. Shine white light onto a very sensitive photodetector (photomultiplier tube or CCD detector), and you can actually count the photons striking

14. From Lakoff and Johnson's book title Metaphors We Live By.

the detector, as predicted by the particle model of light. To further seal the new dual metaphor, two American scientists Germer and Davis managed to diffract electron particles into a series of waves, as if these particles were waves, which they were for this experiment. We ask—would science have ever found the principle of complementarity—the wave-particle duality of matter without first embracing the two metaphors: light is a wave, and light is a particle, first? Can we discover new things without utilizing metaphors?

It is doubtful. However, some metaphors are more apt than others. Good scientists have a way of finding apt metaphors to describe their work and excite the potentiality of their discoveries in the minds of other scientists. But often a metaphor can be chosen that seems "apt" initially, and latter work reveals it has flaws, but like a bad suburban myth on the internet, you just can't get rid of it. Scientific metaphors have lives; they come into existence and persist while they do useful work for us, then slowly die as scientists realize they are no longer apt.

A fascinating scientific metaphor to study as an example is the one created by Watson and Crick of a self-replicating DNA molecule. The phrase *self-replicating DNA molecule* reflects the original understanding of Watson and Crick, and possibly gives clues to their worldview. The metaphor depicts a solitary molecule splitting and self-replicating as it slowly evolves and changes, to give rise to all of life. Nothing else is needed; it is self-contained. The instructions for life are all contained within this one molecule. Obviously, Watson and Crick knew from the onset that this was ridiculous. It is also well known that both Crick and Watson were philosophical materialists. A solution of DNA left on a shelf will do absolutely nothing, no matter how long you leave it there. Without the cellular machinery, search-split-rebind-code proteins, signaling peptides, ribosomes, and so on, DNA does nothing, no matter how much information it contains. Then why was the metaphor *self-replicating* chosen? Did it hide the fact that a chicken-and-egg problem existed? Which came first, DNA or the operating environment, the tape or the tape recorder? Or were Watson and Crick just over-enthusiastic about what the DNA molecule could do?

No matter the cause, as Stephen Rose points out, the result was that researchers were "profoundly misled" by Crick's metaphors describing DNA as self-replicating. Rose suggests instead a symphony metaphor in which the DNA gene is an "active participant" in "cellular orchestra," so is "both the weaver and the pattern it weaves, the choreographer and the dance that is danced."[15]

---

15. Midgley, Science and Poetry.

In Chapter 2—Philosophy First and Chapter 3—The Society of Science, we discussed how the philosophy of science has evolved over the years, but a strong empirical and positivist approach with materialist leanings has persisted to this day. Obviously, many scientists don't feel this way, but a surprising number have persisted in this outlook in spite of its many failings. This empirical legacy traces itself back to the 17th century thinkers John Locke and Thomas Hobbes. Their dream was to construct a world where everything could be described and thought of objectively, so they considered metaphors to be decorative and unnecessarily indirect ways of speaking of things. Thus, Thomas Hobbes concluded, "metaphors are "absurd and misleadingly emotional."[16] Similarly, John Locke stressed, "if we would speak of things as they are, we must allow that . . . all the artificial and figurative application of words . . . are for nothing else but to insinuate wrong ideas, move the passions, and mislead the judgment."[17]

Interestingly, Locke uses a number of metaphors in this criticism of them. He describes passions as things that can be *moved*, and judgments as entities that are *misled*. One might be tempted to think such attitudes have changed. However, the well-known Oxford chemistry professor Peter Atkins makes clear that the empiricist legacy lives on. In his essay, "The Limitless Power of Science, Atkins maintains that, "although poets may aspire to understanding, their talents are more akin to entertaining self-deception. [. . .] They have not contributed much that is novel until after novelty has been discovered by scientists. . . . While poetry titillates and theology obfuscates, science liberates."[18] Atkin's comments reveal a common fallacy in the positivist-empiricist understanding of scientific language—the fallacy that scientific language is free from all subjectivity. In other words, that scientific language is entirely objective. However, as our previous discussion about the metaphors used for light illustrates, human beings choose metaphors based on their best judgment at the time. Sometimes that judgment is partial and incomplete; sometimes it is simply wrong. This fallacy gives rise to what Lakoff and Johnson call the "myth of objectivism," which takes a number of forms in science.[19]

One of the "myths" is that science provides a methodology that allows us to rise above our subjective limitations to provide a correct, definitive account of reality. However, the reality is that scientific theories are developed

16. Hobbes, Thomas, Leviathan, Pt. 1 Chap. 5.

17. Locke, John, An Essay Concerning Human Understanding, Bk. 3 Chap. 10.

18. Midgley, Science and Poetry, 123.

19. Lakoff and Johnson, Metaphors We Live By, 186.

within cultures with particular worldviews, and cultural biases are reflected in scientific metaphors. For example, the scientific term *high-energy* particles; reflects the cultural metaphor, *more is up*. The term *high-level* cognitive functions; uses the cultural metaphor, *rational is up*. To illustrate the inherent cultural basis of our deepest scientific theories, look at the orientation metaphors of time and the future. In English, we describe the future as in front of us. Thus we look *forward* to the future. And we look *back* on our past. But in the Hebrew language, the future is described as behind you, because you can see in front of you, just like we can see our past. But since we cannot see the future, it is as if it is behind us where we cannot see. Upon contemplation, the Hebrew approach makes a lot of sense, but it feels strange to English speakers because the *future is ahead* metaphor is so ingrained. Another example comes from the Nobel Laureate Physics professor Richard Feynman, who was noted for his many contributions to the field of Quantum Electrodynamics. He was an unusually creative and fascinating man with many technical and layman-level physics publications to his credit. He used space-time diagrams to explain the strange quantum mechanical behavior of sub-atomic particles, particularly electron-photon interactions.

Feynman Space-Time Diagrams for Sub-Atomic Particle Interaction

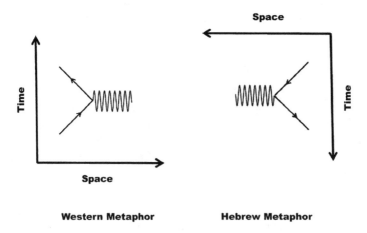

Western Metaphor　　**Hebrew Metaphor**

Shown is a Feynman diagram[20] for the interaction of two sub-atomic particles, an electron and a positron, to produce a photon as a result of the annihilation of the two opposite particles. Look at the diagram on the left

20. Feynmann, QED The Strange Theory of Light and Matter, Chapter 3.

labelled Western metaphor. The arrow going up and to the left is the positron moving forward in space (to the right) and in time (going up). The arrow going up and to the right is an electron that is moving forward in space (to the right) and in time (going up). As we "read" the diagram from left to right and transfer that metaphor to an event occurring, then the electron and positron collide, interact, and annihilate each other to produce a photon that travels on from the particle-particle event to the right. Now since Quantum Electron Dynamics (QED) says that a positron "can" be treated like an electron moving backward in time then we could read the Western Metaphor Feynmann diagram like this. An electron moving backwards in time (down) collides with an electron moving forward in time (up) to annihilate each other to produce a photon moving forward in space (to the right). Now let us imagine the same diagram using the Hebrew metaphor of the future as behind us. Then, "perhaps" his famous Feynmann diagrams would look like the one on the right labelled Hebrew Metaphor. Here the electron and positron move forward in space (to the left) and in time (or down, since the Hebrew language speaks of time as behind) to collide, interact, and annihilate each other to produce a photon that goes forward in space (to the left).

This can get confusing unless we all agree on which metaphors we will use. But one could argue against this up or down, left or right nonsense by saying the change in metaphor is neutral with regard to understanding the scientific concept. If all agree on the orientational metaphors used; the same concept can be used with equal usefulness and understanding in either metaphor. But what if the metaphorical change or shift is not neutral, but limits or expands what can be seen or done with the concept? Then what? It becomes obvious that metaphors are not only used in science; but become critical for its very function. No metaphor, no expansion, no discovery, no science. Language and its accompanying metaphors of all shades and grades are part of how we think. How would our scientific system have arisen if western culture had no provision for mathematical symbolism in our language? For example, not every culture has had a concept or metaphor for zero. Suppose western culture had no concept for zero, or metaphors that communicate it. Feynmann would have never developed his diagrams. Once you see this, it shows how impossible it is to avoid metaphorical language.

A second myth of objectivity is that "words have fixed meanings" and therefore give a final, definitive account of reality.[21] However, as our examples above illustrate, metaphors and models reflect the experience of the one who names the object/concept/action/etc. These change over time and differ

---

21 Lakoff and Johnson, Metaphors We Live By, 187.

between cultures. *All* naming is classification based upon the experience and purposes of the namer, and has no innate correspondence to reality. For example, the three letters that form the word "cup" are arbitrary symbols used to denote our abstraction of a phenomenon that is, in reality, an event not a thing.[22] The physical object we name a "cup" is in reality constantly in flux because of the movements of atoms that make up its substance. The cup you hold in your hand this instant is not the same cup you held just a moment ago. However, to make the world navigable and discussable, we give things names.[23]

A third form of the myth of objectivity is the assertion that to maintain objectivity, metaphor can and should be avoided.[24] However, as shown earlier, science, even cognition, is impossible without metaphors and models. Molecular biologist and theologian Alister McGrath and Australian philosopher J. J. C. Smart respectively emphasize the importance of models in scientific research in the following two statements:

> The basic idea [in science] is to establish a model which is able to explain the most important features of a system, and then develop the model further to incorporate more complex features of the behaviour of the system.[25]
>
> A theory that can be formulated without a model . . . is a dead theory[26]

A model is important for science to advance, and models and metaphors are intimately related.

Lastly, let us dismantle a fourth myth of objectivity: the belief that the metaphors of science are purely rational, while the metaphors of the arts are purely imaginative, and thus less rational.[27] This is an either-or-fallacy. Both scientific and poetic metaphors rely on the imagination as well as reason—on what Lakoff and Johnson call "imaginative rationality."[28] The name "black holes" is a good example. Black holes are black because light that goes past the black holes event horizon never escapes, hence *black* and *hole*.

22. Postman, The End of Education, 180.

23. This paragraph's example and explanation has been adapted from Neil Postman's essay "The Word Weavers, The World Makers," found in Postman, The End of Education, 180–181.

24. Lakoff and Johnson, Metaphors We Live By, 187.

25. McGrath, Science and Religion, 146.

26. Smart quoted in Soskice, Metaphor and Religious Language, 113.

27. Adapted from Lakoff and Johnson, Metaphors We Live By, 187–88.

28. Lakoff and Johnson, Metaphors We Live By, 193.

However, these terms are not literal and, furthermore, they have emotional connotations, just like poetic metaphors do.

Another example is particle "spin." The fundamental particles of matter have a property that is called spin because an electron spinning in a magnetic field was the nearest physical model or metaphor that could be found. But it is widely understood that the electron may not be spinning at all; we just don't know. The standard particle model of physics builds upon the fundamental particle called a quark (though it has never been observed yet). Quarks have spin. They also have other properties classified as "flavors" (up, down, strange, charm, top, and bottom) and other properties called "colors" (red, blue, green). If it seems that the study of nature is taking us into realms where the "lack" of a good metaphor is holding research back, then you are probably right. Quantum chromodynamics expands QED to include protons and neutrons in the interactions of nature. Consider this comment by the physicist Feynmann on the extrapolation of his field of quantum electrodynamics (QED) into the field of quantum chromodynamics.

> The idiot physicists, unable to come up with any wonderful Greek words anymore, call this polarization by the unfortunate name of "color," which has nothing to do with color in the normal sense. At a particular time, a quark can be in one of three conditions, or "colors" —R, G, B (can you guess what they stand for?)[29]

The choice of a metaphor to describe a scientific result can be critical and have far reaching consequences.

But if metaphors are so needed in science, do they still introduce problems, and have limitations? Absolutely yes. *All* metaphors highlight and hide features. *No* metaphor is what it describes, only a hoped-for close approximation. Alister Mcgrath comments,

> Models are . . . not identical with what they model, and must not be treated as if they are. In particular, it must not be assumed that every aspect of the model corresponds to the entity being modelled [sic].[30]

Examples of failed metaphors and models abound in the history of science. The nineteenth century search for the "luminiferous ether," the mysterious medium that light waves "had" to propagate in is one example. But light needs no such medium. Indeed, the wave model of light is only partially correct. Current scientific metaphors may need to be revised or rejected. The decision of a particular scientist to abandon or modify a current

---

29. Feynmann, QED The Strange Theory of Light and Matter, 136.

30. McGrath, Science and Religion, 149.

metaphor is a deeply personal one, based on scientific experiences, and personal tacit knowledge in the field that comes from intense inquiry and self-searching. Repeating a Polanyi quote from chapter three,

> the scientist may appear as a mere truth-finding machine steered by intuitive sensitivity. But this view takes no account of the curious fact that he himself the ultimate judge of what he accepts as true . . . the scientist is detective, policeman, judge, and jury all rolled into one.[31]

So what is a good attitude to take towards the use of metaphors in science? Humility is always a good virtue. Our model may be fantastic and useful, but in time it will likely fail and show many flaws. That is the way of most models. Such failures can stimulate a hunger for the truth, a better metaphor and with it further scientific inquiry.

> [Scientific] worldviews have tended to be accompanied by a prevailing metaphor or analogy. . . . But one intriguing feature of recent discoveries [in quantum physics] is this: the universe they suggest is not like anything with which we are familiar. A universe that allows for instantaneous influences between events that have no connection whatsoever is not a universe that is like anything familiar to us. . . . For the first time in (at least recorded) history, we may be metaphorless.[32]

This quote from Richard Dewitt from *Worldviews: An Introduction to the History and Philosophy of Science* points at a possible future paradox. What if our research leads us to a result that has no metaphor to describe it? What then? Are we confounded and metaphorless, or are we in a place described by an ancient Hebrew prophet?

> Therefore once more I will astound these people with wonder upon wonder; the wisdom of the wise will perish, the intelligence of the intelligent will vanish. *Isaiah* 29:14

Michael Polanyi issued a warning. He saw the danger of imposing a strictly detached and supposedly objective viewpoint on all knowledge. This is inherent in any passionately believed empiricist or positivist outlook on life.

> We can see how an unbridled lucidity can destroy our understanding of complex matters. Scrutinize closely the particulars

---

31. Polanyi, Science, Faith and Society, 38.
32. DeWitt, Worldviews, 305.

of a comprehensive entity and their meaning is effaced, conception of the entity is destroyed[33]

. . . the belief that, since particulars arc more tangible, their knowledge offers a true conception of things is fundamentally mistaken.[34]

At the base of scientific knowledge lies a supporting reservoir of personal or tacit knowledge. This includes metaphors and all of the "art" of good science as passed down by the grand mentoring tradition of graduate schools and post-doctorate appointments. The positivist outlook refuses to acknowledge these personal elements, metaphor construction, sense of research direction, etc., as valid forms of knowledge. Science is a fail-proof method of finding truth, and the sooner we objectivize the methods of science the better. Polanyi then states,

> We are approaching here a crucial question. The declared aim of modern science is to establish a strictly detached, objective knowledge. Any falling short of this ideal is accepted only as a temporary imperfection, which we must aim at eliminating. But suppose that tacit thought forms an indispensable part of all knowledge, then the ideal of eliminating all personal elements of knowledge would, in effect, aim at the destruction of all knowledge. The ideal of exact science would turn out to be fundamentally misleading and possibly a source of devastating fallacies.[35]

Trying to summarize Polanyi; eliminate the personal elements of scientific knowledge, and you have undermined the base on which all scientific knowledge depends. The long term effects on science will be devastating causing it to flounder about chasing petty goals and hypotheses that do not really advance the field or humanity. But if this mentality, coupled with the enthusiasm arising from historically successful science, is applied to other areas of human knowledge: government, law, sociology, economics, etc., the results could be devastating indeed. The intrinsic moral and representative governing structure constructed by western society over the last three thousand years becomes grist mill material for the empirical method; and with it the loss of our accumulated moral base because it cannot be objectivized. Polanyi claimed to have watched this happen in the coffee houses of Budapest before the Great War, World War I. The end result of this Polanyi called a moral inversion. Man is fundamentally a moral creature, he claimed. If

33. Polanyi, The Tacit Dimension, 18.
34. Polanyi, The Tacit Dimension, 19.
35. Polanyi, The Tacit Dimension, 20.

moral sentiments are removed from him because they cannot be objectivized or incorporated in the empirical dream, mankind will find his morals in what he can objectivize, even if that means right is wrong, and a lie is the truth. This is what Polanyi called a "moral inversion," and it manifested itself in the great schemes of Nazism, Fascism, and Communism. Repeating the idea in Polanyi's words,

> The method of disbelieving every proposition which cannot be verified by definitely prescribed operations would destroy all belief in natural science. And it would destroy, in fact, belief in truth and in the love of truth itself which is the condition of all free thought.[36]

I finish this chapter with an ending paragraph from the preface to Polanyi's, *Science Faith and Society.*

> I shall quote here from the writings of Nicolas Gimes, a Hungarian communist who, though he had shortly before been a faithful Stalinist, turned against Stalinism in the Hungarian Revolution of October 1956. The following passage was published three weeks before the revolution.
>
> Slowly we had come to believe, at least with the greater, the dominant part of our consciousness . . . that there are two kinds of truth, that the truth of the Party and the people can be different and can be more important than the objective truth and that truth and political expediency are in fact identical. This is a terrible thought . . . if the criterion of truth is political expediency, then even a lie[37] can be 'true' . . . even a trumped up political trial can be 'true'" . . . And so we arrived at the outlook which infected not only those who thought up the faked political trials but often affected even the victims; the outlook which poisoned our whole public life, penetrated the remotest corners of our thinking, obscured our vision, paralysed our critical faculties and finally rendered many of us incapable of simply sensing or apprehending truth. This is how it was, it is no use denying it.
>
> The author of these lines was executed in Budapest in 1958 at the orders of Moscow.[38]

36. Polanyi, Science, Faith and Society, 76.

37. The original text in Science, Faith and Society contained the wording "if the criterion of truth is political expediency, then even a life can be 'true'". It is the author's opinion in light of the context of this paragraph that life is a typographical error, and should read lie. I took the liberty of correcting this, and accept all responsibility for misquoting if wrong.

38. Polanyi, Science, Faith and Society, 18–19.

8

Change and Evolution

DR. THOMAS MARKLEY CLEARED his throat. "Okay class, today I want to talk about the YEC or young earth creation model. It is the model that I believe along with 2000 years of traditional Christian church history, and much of the Christian population in America. I am not even mentioning the thousands of years of Jewish history wrapped up with this model."

It was an interesting class he was speaking to; Bert, Jon, Ellie, Charles, Sachie, Mary, Tolu, and about a dozen or more students. "Not everybody in my scientific discipline agrees with me, both religious and not. In fact, very, very few in my field agree with me; but that is another story. I have very good reasons on why I believe the way I do, and now I am going to lay out my case."

Bert whispered to Sachie, "Does Dr. Markley know Dr. Delzer?

"Who knows," whispered Sachie back.

Markley glanced over the class again, "Young earth creationists (YEC) agree that the creation days listed in Genesis chapter one were six literal 24 hour days and they happened somewhere between 6,000 and 12,000 years ago. We also believe that around 2,300—3,300 years before Christ, there was a global flood (Noah's flood) that radically re-arranged the surface of the earth. Just as the Biblical narrative states, all of the land animals, birds not on the ark, and some of the sea creatures were buried in the sediments. Thus the global flood was responsible for most (but not all) of the fossilized remains and geologic layers that we see on the earth's surface. Some fossils may have been produced before the flood, but generally what we see, including the segmentation of the species by rock layer, is a result of localized catastrophic sedimentation events that occurred during the Noahian flood."

Dr. Markley was a YEC oriented earth scientist. He and Dr. Delzer had known each other for decades. "Genesis chapter 1 is history, it's not a parable or a poem or a prophetic vision. Look at the verbs used in Genesis 1 and the way Jesus refers back to Genesis 1. Note also how the Hebrew word for day, *yom*, is used in Genesis 1. It always has an ordinal number in front; one day, second day, third day, and with evening and morning. Everywhere else in the Old Testament when a number precedes a day, it means a literal 24 hour day. If you look at the chronology in Genesis, that is the order of creation. This is a bit at odds with the secular geologic record. The earth was created before the sun and stars, fruit trees were created before any sea creature, and birds were created before dinosaurs. Note that the Jewish work week matches the creation week; God did not say "work six days because I created six long, indefinite periods." If God created over millions of years, he would not have a reason for Sabbath keeping." Dr. Markley glanced around and then asked, "Questions at this point?"

Bert's hand shot up, "Dr. Markley, accurate literal reading of the Hebrew in the Old Testament has the sun coming out of its house from beneath the earth (which was flat), and then traversing the sky during the day to go back into its house underneath the earth. This is obviously geocentric. Do we have to believe that to properly believe the Genesis account?"

"Ah, good question, Bert." Dr. Markley was not surprised at the question, nor was he keen about it. "The early Hebrews obviously read Genesis that way but I am not sure that is what the Bible really says. In other places in Genesis it talks about the circle of the earth, like Isaiah 40:22 New Living Translation—"He sits enthroned above the *circle of the earth*, and its people are like grasshoppers. He stretches out the heavens like a canopy, and spreads them out like a tent to live in." It is very true that the ancient Hebrews may have looked at this passage as a flat circular earth, but it does not mean that we have to do likewise. Let me finish out what I have to say first."

Bert nodded assent and Dr. Markley continued. "If we buy the miracles of Jesus then why can't we buy the miracles of God in creating the world in the first place? Is that too much to ask? Jesus taught that man was created in the beginning, see Mark 10:6.

Pausing to gather some verbal strength Dr. Markley started. The main problem with an old earth view or OEC; is that it has trouble in dealing with the nature of sin and evil. How did evil enter a perfect world? How did we get a sin nature that Christians everywhere believe we need to be delivered from? The Bible teaches us that prior to the fall of Adam and Eve in the Garden, when they ate the forbidden fruit, the world was "good," perfect. There was no sin, no death and no carnivorous behavior by animals or man. To stretch the days of creation into indefinite long periods of time, that created the fossils that we see in the geologic layers, is to imply an enormous amount of death and suffering on the part of the animals that lived and died in these times. And all of this occurred before the Fall? How do you justify this?

"But how do you justify a mile thick layer of flies that never dies!? Thousands of microbes that die, crunched underneath Adam's foot every time he walks across the garden? Poor dead plants every time Eve eats dinner as a vegetarian?!" Bert was completely exasperated.

Markley eyed Bert impatiently, but restrained himself before replying. "Bert, don't cut off a speaker when he is expounding his points, give them a chance to fully develop their ideas before you start to disagree and attack. You might actually learn something and change how you reply or question the speaker. You cut me off, but let's stop and answer your questions as you may not be the only one with these questions."

"I am sorry, that was rude of me, but you have created a world that goes haywire when you eliminate all forms of death." Bert looked a little cornered, but was not in a backing down mood. "At the rate that flies reproduce, if they don't die we would be covered with them in a few months to years. The same with microbes, the cells in our body that have to die every few days or weeks or else we develop cancer. Death is intrinsic to our very world and ecological balance. How can it be all bad?" Tolu and Katrina looked at each other with a "this is getting bad" sort of look.

Tolu leaned over to Katrina and whispered, "The only thing that could make this worst is for Dr. Delzer to show up, and start throwing his views into the picture."

"Bert, calm down, I never said that all death came with the fall, just animal death. Adam and Eve did not die right away after they sinned, they started the processes by which they eventually died; so "death" came into their lives. The pre-fall world did not run like our "fallen" world does today. Thus problems with crushed microbes, or over-populating flies and microbes simply do not apply. They did not need to reproduce like they do today to survive. The rules of creation and living and ecological balance were just different. Don't ask me how; I just know it does because the Bible and Genesis imply that it was; it had to be. Dr. Markley wiped his brow and continued. "Bert, you are a philosophy major, right"

"Yes sir."

"Different viewpoints are the fodder of philosophy"

"Absolutely."

"Then it is imperative that different viewpoints be encouraged as long as you can appropriately justify them philosophically and logically"

"Of course."

"Otherwise you would have nothing to discuss or philosophize about, and the advancement of new ideas in the field would come to a halt."

"I think I see your point," said a nervous Bert. "I just wanted to debate your points."

"Of course," said Dr. Markley. "And that is fantastic as long as you give your opposing viewpoint adequate time to develop their ideas and they are open to constructive criticism and debate in the first place."

Scanning the class, Markley could see the tension had been relieved; now it was time to continue his saga. "As you can see if you are familiar with the Genesis account of creation, and I realize that not everyone in this class is, that I am putting the plain sense reading of Genesis one on a higher plane of truth than secular science. It is a matter of who do you take as your higher authority, geology or Genesis one. It is not that geology or science is bad or evil, for heaven's sake, I am a scientist! It is just that secular geology started with some wrong fundamental assumptions and then started to interpret the geologic record. With a different set of assumptions you can interpret the geologic data differently and see a young earth with layers and fossils deposited by a worldwide catastrophic flood. It doesn't mean we have all the answers figured out or can operate as tightly and neatly as the secular version of geology. If we had the resources, monies, and people that secular geology has where would flood geology be by now? Maybe we would have all the answers."

Slowly the classroom side door opened and in stepped Dr. Delzer. "Oh, sorry Tom, I thought you had finished class; so sorry to intrude."

"That's fine, I have just finished up anyway," said a smiling Dr. Markley.

"Are we rendezvousing at the same place with our campers and families as last month?" Dr. Delzer was turning to step back out of the classroom.

"As always, we can't miss the monthly camping trip with our families. Who is going to get the biggest bass this time?" replied Markley.

"My son probably, he is passing both of us these days"

"Have you finished proofreading my paper on young earth creationism and Kant yet?" asked Dr. Markley.

"You know I don't agree with any of it, you know."

"Of course, that is why I gave it to you to proofread, who else but a friend is going to catch the weak spots."

"Will be my honor, have it by this afternoon when we meet up, see you then,"

"See you then," replied Markley. With that Delzer was out of the room and down the hallway.

Ellie was looking very puzzled. Bert was picking his jaw up off of the floor. "But isn't old earth creationism compromising on the Word of God.

Introducing a fatal flaw into the entire scheme of Christian redemption?" queried a very confused Ellie to Dr. Markley.

"I refuse to be a "Jesus and" sort of Christian. It is always Jesus and this; Jesus and that; Jesus plus this. This is the starting place for cults and heresies. It is Jesus only. No Jesus plus this or that. Two thousand years of orthodox Christianity has been united on this point. We have divided on every other point maybe, but on the singular saving death and resurrection of Christ and our acceptance of it and Him; not on our opinion on the age of the earth, or the how of the fall of humanity. Geez, I dislike these "Jesus and" legalists that demand you find Christ through their lens and their lens only. Class dismissed, got to go." With that, Dr. Markley picked up his notes and left quietly.

Douglas Axe is a molecular biologist who spent a decade working at the premier molecular biology lab in the world, or at least, one of the premier labs. He started his career as a chemical engineer who had an "aha" moment when he first studied biochemistry. Biochemistry is crawling with cycles of linked chemical reactions and mechanisms that produce the energy, proteins, and raw materials our bodies need to function. In most of these cycles and chemical cascades of biochemical reactions are innumerable feedback loops where the product of one reaction is used to control or moderate one of the precursors that created it—a feedback loop. Feedback loops whether chemical, electrical or mechanical are very complicated systems to design and manage well. Entire university courses are dedicated to understanding and controlling them. Their presence in naturally occurring systems was a shock; that begged the question of where they come from, how they originate—or more importantly, how were they "invented." Axe changed his career to molecular biology; in particular, protein research. His story is the stuff of legend; in scientific findings, and horror when your findings buck the accepted consensus. It is a classic study in philosophy of science, and the human nature of science. The saga is recorded in his book *Undeniable*[1].

Axe realized that ultimately evolutionary theory was based on the ability of living organisms to "mutate" and then select or "evolve to" new proteins from earlier precursor proteins. Since proteins are the building blocks of life, new proteins are arranged to give new tissues that give new organs and muscular systems that allow one species of organism to evolve into another species, and eventually even further. If natural selection and evolution cannot evolve new proteins then the entire theory of evolution

---

1. Axe, Undeniable.

is in danger. But let us define some concepts to fully understand what is at stake in this question.

The general concept of evolution, that one type of animal comes from another goes back to the Greeks. David Barton[2] writes of this early history with references[3] as follows:

Anaximander (600 BC), introduced the theory of spontaneous generation; Diogenes (550 BC), introduced the concept of the primordial slime; Empedocles (495–455 BC), introduced the theory of the survival of the fittest and of natural selection; Democritus (460–370 BC), advocated the mutability, and adaptation of species; the writings of Lucretious, before the birth of Christ, announced that all life sprang from "mother earth" rather than from any specific deity; Bruno (1548–1600), published works arguing against creation, and for evolution in 1584–85; Leibnitz (1646–1716), taught the theory of intermedial species; Buffon taught that man was a quadruped ascended from the apes, and Erasmus Darwin, (Charles Darwin's grandfather), discussed evolution in his book Zoonomia (or the Laws of Organic Life, 1794).

Charles Darwin, in the first 1859 edition and then later editions of *Origin of Species,* introduced natural selection as the mechanism for accomplishing the gradual change in species from generation to generation until a new species was formed. Variants are often produced in new offspring, and Darwin proposed natural selection as the agent that "selected" the variant as a new base of species if the variation offered something to enhance the reproduction or survival against disease and predation of that variant. Then its descendants would take over the species population. The variant passed on its beneficial variations to its descendants who eventually "become the new normal" for the species. With enough time the variations could build up to reproductively isolate the new species from the old species. A new species was born. With vast quantities of time these species changes could add up to new genus, new kinds, new families, and even new phyla of animals and vegetation. This was Darwin's great, gradually developing *tree of life.* From one or a few simple forms of life came everything. The origin of the first species he did not even speculate on until his famous letter to Joseph Hooker in 1871, long after the first publication of *Origin of Species.* Axe discusses in *Undeniable* how Darwin fought rejection of his idea, from 1859 until the 1872 sixth edition of *Origin of Species.* That is when the naturalist community seemed to change and Darwin's theory was almost universally accepted.

By the start of the twenty century, Darwin's theory of evolution by means of natural variation and natural selection was in serious trouble. It

---

2. Barton, "A Death Struggle between Two Civilizations," 3–56.
3. Clodd, Pioneers of Evolution from Thales to Huxley.

was becoming apparent that, natural variation in species was not up to the task of creating new species. No one had ever bred a pigeon into another type of bird, a cat into a dog. By 1930–1940 several leading biologists had fused Mendelian genetics with natural selection to form the neo-Darwinian Synthesis. Four forces operate on populations to cause evolution: random genetic drift, gene flow, mutation pressure, and natural selection. But the key to remember is that natural selection is the factor (survival of the fittest) that selects out of these forces what will get passed on to future generations. But what generates the significant enough variation to get an organism past the species limit? The Neo-Darwinists postulated that mutation pressure and in particular random genetic drift in the genome would do the job. By the fifties and sixties it was apparent that chemical, radiation, or copying errors in the replication of DNA was the source of this genetic drift that eventually lead to new species, etc. The key determining factor was *random*; or *stochastic* as mathematicians like to say. The copying errors were not *intentionally* introduced. There was no goal or plan behind them. They were *random*. Thus the generation of new plants and animals did not require any "intelligent" input but could serve as materialistic explanation for life on earth. At this point two perspectives developed on how the random changes in the genome were fixed in the new species. The molecular view says that natural selection "selects" and fixes the changes most beneficial to the species. The neutral view says that the random drift just happens and the set of errors that happens to dominate becomes the new species and the organism moves on. In both approaches you still must have the "accidental" changes in the genome that the organism "drifts" to or is "natural selection" selected to.

Not every biologist sees this as a totally materialistic process. Some Christian and theistic oriented biologists claim that God was active in the set-up of the process and uses randomness as a design agent. I have even seen books written with unusual titles such as *Random Designer—Created from Chaos to Connect with the Creator*[4]. Most materialists find this unnecessary and believe in chance and necessity alone. If random chance is sufficient to create, innovate new organisms then why do we need to bother with some kind of intelligent designer like God? God using a random process, (that doesn't need Him), as evidence of His creative power is not a very persuasive argument for the materialist.

Douglas Axe saw another possibility. Was it possible to use the current state of protein research to test whether or not random chance genomic variations have what it takes to develop a new fully functional protein? If so then, chance and necessity are possibly sufficient to develop new proteins.

4. Collings, Random Designer.

If not, then they can never do it by random chance alone. To understand what is at stake; look at the Evolution—DNA to Organism figure. Under the Neo-Darwinian model we could simplify the steps of evolution as shown in this diagram.

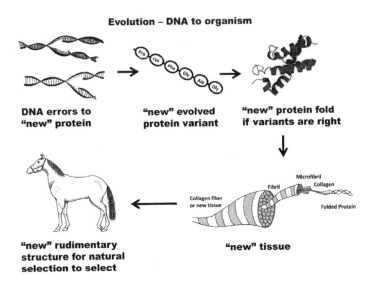

Evolution – DNA to organism

DNA errors to "new" protein

"new" evolved protein variant

"new" protein fold if variants are right

Microfibril
Fibril   Collagen
Collagen fiber or new tissue
Folded Protein

"new" tissue

"new" rudimentary structure for natural selection to select

Copying, transfer, or other errors in DNA, the transcription process, or the assembly of proteins creates a new slightly different protein or a variant protein. If enough "point mutations" (change in one of the amino acids in the primary sequence of the protein) are accumulated and preserved (by natural selection or otherwise) then a new protein fold may develop in the protein to help create a new functional structure of that protein. That "enhanced" or "new" function causes or creates a new or altered tissue that develops into a "new" or "enhanced" structure in the organism that natural selection can select by allowing the organism to survive better, fight against predators better, reproduce better, etc. This gives the organism the reproductive edge so that its offspring will out populate the rest of the population, and hence become dominate in that population after many generations. Thus the new structure becomes the new normal for that species.

A protein is "functional" if it carries out some biochemical task or reaction. That requires that the protein fold into a proper shape or conformer. Many proteins (peptides) will not fold. Many will fold into structures, but these will not do a specified task or any biochemical reaction or task. They are not "functional." Dr. Axe knew that the error rate of point mutations (change of an amino acid in the primary sequence) in proteins per generations of a species, and the size of a species' population was fairly well known. Also the ability to form a new functional protein is not at all guaranteed every time a new protein variant is formed. Axe wanted to find out the odds that a new or enhanced functional protein would be produced for every point mutation or protein variant produced. If the odds were high enough then random mutations had a chance to produce a new useful protein with every new protein mutation, and natural selection would not filter that protein variation out. If the odds were very low then the new variant protein produced would be eliminated by evolution because it contributed nothing to enhance the survival or reproduction of the organism. It is well known that organism structures and the proteins they are built of cost the organism energy and resources. If structures or proteins do not do something to enhance their survival or reproduction (survival of the fittest), the organisms with this drawback will be eliminated from the gene-pool, and eliminated from the species. Dr. Axe saw this phenomenon in action with his research on the *beta-lactamase* gene[5]. This bacterial gene encodes for the protein (enzyme) *beta-lactamase*. *Beta-lactamase* inactivates the penicillin molecule so that it does not kill the bacteria. This enzyme has a cleft built into the folded structure that chemically changes and deactivates the penicillin molecule. Axe and his group mutated (changed the primary sequence coded in the gene) so that the beta-lactamase barely worked; just enough to save the bacteria from the deadly penicillin molecule at very low doses. They then allowed natural selection through six rounds of mutation and selection to repair this faulty gene. Sure enough at the end, the beta-lactamase produced was repaired and functioning at the original level or slightly better than the original beta-lactamase. In this case Axe had left the critical cleft region of the *beta-lactamase* basically alone and had mutated the rest of the molecule. The invention of the cleft was kept intact and the rest of the protein machine was fine-tuned by natural selection for optimal operation. (Machine, machines-why does the term and descriptor machine keep popping up-shades of Michael Polanyi?) In the next experiment Axe and his group deleted 108 DNA coded bases including those that produced the famous cleft region in *beta-lactamase* that inactivated penicillin. In spite of this serious mutation, the mutated *beta-lactamase* still had the ability to inactivate the penicillin molecule even if weakly, but definitely not by the cleft mechanism[6,7] of the original *beta-lactamase*. After several rounds of mutation and selection, natural selection was unable to improve

5. Axe, Undeniable, 104–7.

6. Axe, "Model and Laboratory Demonstrations," 1–13.

7. Axe, "Estimating the Prevalence," 1295–315.

on this weakly functioning enzyme or reinvent the critical cleft region of the original enzyme. This gives us a striking picture of natural selection at its most fundamental level, the level it *must* work at to achieve its lofty claims. Natural selection cannot *invent*, it can only *fine tune* what has already been invented by other means. At a larger level we could ask how does natural selection select for something that does not yet *exist*?

Dr. Axe was fascinated by comments made by Michael Denton and other critiques of evolution that appeared in the 1980's. Dr. Michael Denton published a very influential book named, *Evolution: A Theory in Crisis*[8], in 1985, where he quoted,

> There are, in fact, both theoretical and empirical grounds for believing that the a priori rules which govern function in an amino acid sequence are relatively stringent. If this is the case . . . it would mean that functional proteins could well be exceedingly rare . . . As it can easily be shown that no more than 10^{40} [1 followed by 40 zeroes] possible proteins could have ever existed on earth since its formation, this means that, if protein functions reside in sequences any less probable than one in 10^{40}, it becomes increasingly unlikely that any functional proteins could ever have been discovered by chance on earth.[9]

Axe wanted to put an actual experimental probability on the odds of finding a functional protein by blind chance or random seeking; and see if it was higher than 1 in 10^{40} or much lower than 1 in 10^{40}. By 2004 he was at the Laboratory for Molecular Biology at Cambridge University, UK and doing the necessary experiments to find a good approximation to that probability. What he found[10] startled everyone who was familiar with the problem. It was not 1 in 10^{40} but 1 in 10^{74}. What does this mean? For every possible "good" functioning protein there are 10^{74} "bad" ones that don't fold or otherwise do anything significant or "functional." Or for every one good functional protein that natural selection can select for; there are 100,000,0 00,000,000,000,000,000,000,000,000,000,000,000,000,000,000,000,000,000, 000,000,000,000,000 proteins that are not functional that natural selection cannot select for since they have no function or advantage to add to the organism. In fact, natural selection will likely inactivate the mutated genes producing these "bad" proteins because of the cost to the organism. Well then, if only 10^{40} proteins have ever existed on the earth, then how could

---

8. Denton, Evolution: A Theory in Crisis.
9. Denton, Evolution: A Theory in Crisis, 323.
10. Axe, "Estimating the Prevalence," 1295–315.

ANY functional protein have been found by random chance driven natural selection?

What does this do to random chance driven evolution if you take the data seriously? It destroys it. Time to move on to another theory of origins. However as our earlier chapters suggest and as Axe found out, (when his paper was published in the Journal of Molecular Biology in 2004); scientific consensus does not change that easily. A single experiment does not change a field over-night, especially when so much philosophical outlook, (materialism, other), time, money, and prestige is wrapped up in in maintaining the status quo. It is easier to ignore the dissenting facts, and go on one's merry way. It would be unfair to charge all of science with this criticism, but in the field molecular biology it may be merited.

Charles was trying to explain the idea of spaces to Ellie and Katrina. Mary, Jon and Bert were sitting in the back of the room causally listening in. Mary was leaning up against Bert, playing with his fairly short hair. Jon ignoring her was intently watching Ellie and Charles.

"What in the world do you mean by a space? You obviously don't mean a bunch of empty area," Ellie was looking a bit desperate as she asked and Katrina was looking a bit amused.

"Okay, slide this book along the floor in any direction," Charles calmly kicked one of Mary's textbook across the smooth dusty linoleum floor.

"Hey!! That's my book!" screamed Mary.

"Oh, I'm sorry I just wasn't thinking."

"I'll say!" murmured a perturbed Mary as she resumed fiddling with Bert's hair.

"Well . . . so sorry . . . As I was saying," Charles a bit red-faced resumed his lecture, "You can describe the line of travel of the sliding book with a vector or line on the floor. You could then represent that line of travel or vector on the floor as a combination of two other lines or vectors set up along the sides of the floor, i.e. the x direction, and the y direction. You resolve the line of travel into its x and y components. In fact, any straight line you draw on the floor can be resolved into its appropriate x and y components."

"Okay, what the heck is a space then?" Ellie was asking.

"I'm getting there. The sides of the floor, the x component of travel and the other side (at a ninety degree angle to the x side), the y component of travel are the *basis* for the floor area. The floor area is the *space* under which the book can experience all possible sliding motions. And the *basis* the x and y components *span* that *space*."

"You are losing me Charles."

"Hang on, I'm not done yet"

"I am," said Katrina, with a mischievous look in her eyes.

Rolling his eyes, Charles continued, "Okay now expand this to a three dimensional room." With that as a cue, Mary grabbed a pen out of startled Bert's front pocket and flipped it across the room. Not missing a beat Charles continued, "Now in the three dimensional space of this room, we can represent any motion of Bert's pen as a combination of the sides of this room, the x and y and z components of the pen's motion. The x and y and z axis (Cartesian axes) are the basis of the space, where the space is the volume of this room."

"Impressive," murmured Bert.

"Now," Charles looked as if he about to give the concluding remarks for a Nobel Prize speech. "Let's expand the room from three to five dimensions, and we could represent the travel of a thrown pen in five dimensional space easily using the x, y, z, t, and u sides of the room. The five dimensional space is spanned by the five basis x, y, z, t, u."

"Bravo! Now I am completely lost!" exclaimed Katrina.

"That's your problem," whispered Charles. Mary was giggling. Bert was oblivious, lost in hair twiddling rapture.

Ellie looked at Charles, "What does this have to do with Douglas Axe's protein research that you were talking about?"

"Take a three amino acid peptide (very small protein). How many possibilities have we got for that protein?"

Mary got up; leaving a surprised Bert behind, went to the board and quickly scribbled.

$$20 \text{ possible amino acids}^{\text{\# of amino acids}} \text{ or}$$

$$20^3 = 8000 \text{ different possibilities for this protein}$$

Charles intoned, "8000 is the space or range of possibilities for that protein." Charles went up to the board as Mary returned to Bert, and gathered her books together. "Now let's take the protein with which Axe was experimenting. It was 153 amino acids long. Its range of possibilities or protein space was

$$20^{153} = 1.141 \times 10^{199} \text{ different possibilities for this protein.}$$

This range of possibilities is so huge as to be beyond the ability of our minds to visualize it."

"But what if 10 percent of these proteins were functional?" asked Ellie.

"Then you would have

0.1 X 20^{153} = 1.141 X 10^{198} functional possibilities for this protein.

"What about 1 percent of these proteins"
"Then . . ."

0.01 X 20^{153} = 1.141 X 10^{197} functional possibilities for this protein.

"What about 0.000001 percent of these proteins"
"Well then . . ."

0.000001 X 20^{153} = 1.141 X 10^{193} functional possibilities for this protein.

Ellie paused and asked, "Well what approximate number of functional possibilities did Dr. Axe find for this protein?"

Charlies quickly scratched on the dry erase board with his marker

Approximately 1.141 X 10^{125} functional possibilities for this protein.

"That gives us the following . . ."

$$\frac{1.141 \times 10^{125} \text{ functional protein possibilities}}{1.141 \times 10^{199} \text{ all protein possibilities}} = 10^{-74}$$

"Or 1 in 10^{74} possible proteins is a functional one," Charles stepped back and looked quite pleased with himself.

"Is this going to be generally true for all proteins of this length?" Ellie queried.

"Yes," he replied.

"Give me a visual picture of this so I can wrap my head around this number."

"Okay, I will try, but it will probably be far too generous and underestimate the severity of the situation."

"Give it a try."

Charles swallowed and thought a bit, Mary had left the room but Bert was still there in the back. "Okay, pretend natural selection, which operates like a blind search, is like a drunken blind astronaut trying to make his way back to Cape Canaveral, Florida to the Apollo moon landing rocket. Every step he takes is a mutational error accumulated in a generation of an organism. Every loud noise attracts him and he wanders blindly in the general direction of the noise, but every time he runs into something he careens off of it blindly in another direction, just slightly favoring the direction of the loud noise. That is natural selection "selecting or favoring" that mutational change because of the functional advantage it gives the organism. He is blind, half deaf, and almost stone drunk. Now drop him somewhere west of Fort Worth, Texas and let him wander. What is going to happen?"

"He will never make it."

"Why?" replied Charles.

Ellie paused, looked up at the ceiling, and back down again, "He will never get past the noise of the Dallas-Fort Worth International Airport, or the nearest freeway."

"You got it! Likewise natural selection will never get past its nearest functional protein to the protein it needs to get to next; the next correct functional protein that is so incredibly far away in protein space and so, so, so rare. Blind searches cannot invent or create. They just home in on the closest noise and get stuck there."

Katrina spoke up, "But what if the astronaut had a voice recorder with instructions recorded on it. Like—take three hundred steps towards the east—go up the stairs—pay money for a bus ticket to Atlanta, Georgia, etc. Then he could make it."

Bert stood up in the back and walked forward to join Charles and answered for him, "Yes that would work but you don't have a blind search anymore, you have intentionality, a goal directed search. Philosophically, you have introduced what we call teleology into the search. It is not random anymore. It shows the marks or signs of intelligent direction or intervention in the blind search."

Katrina looked puzzled, "But isn't that what theistic evolution is all about; God helping evolution along the places where random chance and necessity cannot do it? I have even heard some Christian biologists and scientists call it evolutionary creationism."

Bert looked a bit aghast, "Yes and no. I wish life was that simple. But there is a lot of term switching and misunderstanding going on with these terms—theistic evolution and evolutionary creationism. Remember the many definitions that can be assigned to the word evolution. Evolution can mean:"

1. Change with time

2. Change of species and genus with time—(some call this microevolution)

3. Nonliving pre-biotic soup to a living replicating cell

4. Generation of all orders and families of life from simple life forms (some call this macroevolution)

5. Generation of man from animals

6. Formation of the galaxies from hydrogen and helium following the Big Bang

7. Formation of our solar system from the matter surrounding our sun

8. Generation of all forms of life from simple life forms by teleological means

9. Generation of all life from inorganic matter by means of random or stochastic processes

At this point Bert was busy scribbling the list on the board to everyone else's staring gaze. "And there are more definitions I could write; this is just a starting point."

"So what is your point, Bert?" Ellie quizzed.

"In the academic and scientific world when you are talking about the evolution of life definition 9, by means of randomness, with no goal or teleology is the only accepted definition that you can talk about in the *scientific* literature. Definition 9 is what is generally accepted as true in the academy. Any experimental result tipping toward Definition 8 can get you blackballed from the standard academic journals as Douglas Axe found out. Thus theistic evolution really ought to be called theistic change with time, or theistic teleological evolution if that is what you really mean. And evolutionary creationism should be labelled as changing creationism or teleological evolutionary creationism. But the terms are almost never described in that fashion but almost always as, how God uses random stochastic evolution to create our world. But all good materialists know if it truly is a stochastic material driven process, you really don't need God; but we don't have to tell the theists that, just get them thinking our way at first."

"My God, Bert; you are pretty touchy on that point!" exclaimed Katrina.

"I guess I am, but when you first encounter these terms theistic evolution or evolutionary creationism, it is very natural to think that you are blending teleology or intelligent driven intentionality with the current scientific model of evolution; a nice happy compromise between two competing paradigms. And many scientists and laymen sincerely believe this way and fall somewhere on the spectrum of God's or the intelligent designer's degree of intervention in the partially random driven process. But if you go to some of the main advocates of theistic evolution or evolutionary creation you really get something else. Evolution is a random stochastic process with no goal or intentionally behind it, because that is what evolutionary biology claims the evidence supports. However we theists know that God used this random stochastic evolution to create the world and all of life. But he has done it in such a way that we will never be able to scientifically see his guiding hand," Bert sat down in a nearby chair.

Katrina looked at him intently and asked, "You obviously don't like this approach, why?"

"Lots of reasons, let me give you just a few. One, it has the philosophical convincing, standing power of a wet rag. For example, suppose I tell you that green gremlins hold me up every time I walk, you cannot see them, there is no scientific test or observation that suggest that they exist, but because everybody in my fellowship has occasionally seen and sincerely believes these green gremlins are holding each one of us up when we walk, then we all know that they are real and really responsible for each one of us being able to walk. And if it wasn't for those green gremlins you my non-believing friend would not be able to walk either. Convinced that green gremlins help us walk?"

"No, not at all," Katrina was laughing at this point.

"Neither am I," exclaimed Bert. "These theistic evolutionary viewpoints strike me as disingenuous, or bait and switch. At first glance they promise one thing, genuine harmony between science and religion, but in reality give you green gremlins. No one on the other side, the materialists, is persuaded to change their minds. They just snicker behind the theist's back. What about all the arguments on cosmic fine tuning, chemical evolution, molecular evolution, etc. Why would the materialist get so upset over them unless it raised very serious threats to their viewpoint?

"So what you are telling me Bert," Katrina was raising her finger lightly in his direction. "Is that the materialists and marginal theists are covering up their weak spots by providing a defensive out for theists (theistic evolution) that handicaps them from ever making gains in the scientific world against the materialists. It also sets up their youth as easy prey for a materialist conversion in their formative years when bombarded by extensive materialist arguments."

"Not all theistic evolutionists think this way, but I wonder about the leaders of this philosophical outlook and opinion. Green gremlins are poor defense against four years of collegiate materialism in my opinion. You have to go for the throat, the basic assumptions from which materialists and agnostics operate. A classic example is the God of the Gap argument that is so frequently thrown at Christians and other theists. God is in the gaps, the unknowns of science, and as science finds out more and more about our world then the need for God to explain anything goes away as the gaps disappear. The tendency is to say yes, but what about this gap that will never fall away, etc., etc. Rather we should be asking; materialism has made a lot of assumptions about its future ability to understand everything and its ability to deliver on its promises. Is this really true? The God of the Gap argument cuts both ways, we need to examine Materialism of the Gaps, and the ridiculous idea that God cannot be proven by science. Neither can materialism be proven by science. And does science prove anything at all? Or does it just build up a case by *modus tollens* logic for whatever you are examining: be it materialistically oriented or theologically oriented?"

Katrina looked concerned, "Bert, you seem to think that science and religion are constantly at war with other, and one or the other has to win."

"No, no, that is not my perspective, it is much more nuanced."

"What do you mean?"

"There are four general ways you can visualize the interaction of science and religion. Sort of like a crude map with some ebb and flow to it. One is the conflict model. Science and religion are in constant conflict with each other. They are at war. Alister McGrath talks about this model and the three other models in his book *Science and Religion*[11]. For example, the biologist, Richard Dawkins argued that, "faith is one of the world's great evils, comparable to the smallpox virus but harder to eradicate[12]. The creationist Henry Morris published a critique of evolution entitled *The Long War Against God* (1989)[13]. Another model is the independence model. Science and religion speak to separate realms, and hence do not really overlap. Some argue that science is concerned with asking "how" questions, and religion asks "why" questions. In fact in a 1981 policy statement, the American National Academy of Science declared,

> Religion and science are separate and mutually exclusive realms of human thought whose presentation in the same context leads to misunderstanding of both scientific theory and religious belief[14].

"Do you buy that?" asked Katrina.

"Of course not," replied Bert. "Life and reality are not nice neat compartments into which we can put everything. You just cannot separate how and why so neatly, the two are often badly interlocked particularly in science and faith issues.

"And the other two models?

Bert started elaborating again, "They are the dialogue model and the integration model. In the dialogue model science and religion are engaged

11. McGrath, Science and Religion—A New Introduction, Chapter 6.
12. McGrath, Science and Religion—A New Introduction, 46.
13. McGrath, Science and Religion—A New Introduction, 46.
14. McGrath, Science and Religion—A New Introduction, 47.

in a dialogue that can lead to greater mutual understanding. For example religion and science both use personal judgment in creating theologies and scientific models, and both use data that is very theory laden. The integration model seeks to merge the two realms into a unified whole that tells one tale of the universe and its origins."

"Well, which do you believe?" Katrina queried.

"All of them and more, life is complicated. Right now in my replies regarding Douglas Axe's research and theistic evolution in the public square, I am dwelling in a warfare mode mostly. But I could go to a few universities and talk about Axe's research and its implications for science and find myself working out of a dialogue or integration model."

Katrina started for a moment, "Are there such universities?"

Bert replied, "There are a few, but not as many as you would think."

With that Charles strolled over and said, "I am tired of thinking, how about some pizza at Som and Ela's Chicken Palace?" They all nodded in agreement and filed out in search of the perfect pizza.

Now is a good time to introduce another scientist who is quite famous in the intelligent design movement and the origins of life debate, Professor Michael Behe. Dr. Behe is a professor of Biological Science at Lehigh University, Bethlehem, Pennsylvania. He is a biochemist with over 40 technical publications, and studied DNA at the National Institutes of Health in his early years. He is best known for his 1996 book, *Darwin's Black Box: The Biochemical Challenge to Evolution*[15]. The book became a bestseller, making the NY Times bestseller list, and ignited a firestorm of controversy. We will start with some of his later more complicated research, and finish up with his key concept *irreducible complexity*, as discussed in *Darwin's Black Box*. If you have trouble understanding Behe's later ideas, hang tight, irreducible complexity is pretty easy to grasp by everyone, and sums up in a nice nontechnical package, much of the molecular biochemical case against evolution that we have been discussing.

First let's start with a paper written by Michael Behe and David Snoke[16], *Simulating evolution by gene duplication of protein features that require multiple amino acid residues* in the journal *Protein Science*. In this paper Behe and Snoke develop a model for predicting how long it would take for random chance driven evolution to produce two point mutations (change of two amino acids in the primary sequence of a protein) in the gene that produced that protein given a known and commonly accepted

---

15. Behe, Darwin's Black Box.

16. Behe, "Simulating evolution by gene duplication," 2651–2664.

mutation rate per generation of an organism and a given population size of that organism. In short Behe and Snoke are asking the question: can chance driven evolution produce enough mutation changes in an organism's genome to explain the changes seen in that organism, over the history of that organism's existence on earth. What makes this fascinating is the answer they found; and particularly the reaction they got from the journal's readers. The story of the response comes from a verbal talk given by Dr. David Snoke at the American Scientific Affiliation Annual Meeting[17] held at Messiah College, Harrisburg, Pennsylvania. I the author was present at that talk and heard this first hand.

Gene duplication has been suggested as a common means of producing evolutionary change. Take a string of DNA, that codes for a particular protein or series of proteins that is needed by the organism. During replication of that piece of DNA, a replication error occurs that causes that "gene" or necessary strip of DNA to be copied twice in the new DNA strand. We now have "two" genes or duplicate genes. One gene is used often by the organism and continues to code out the necessary proteins, etc. for the organism's survival. The other duplicate gene is not used much and so accumulates point mutation errors (change of an amino acid in the gene coding) during multiple replications. A mathematical model can be built from a known mutational error rate, the number of replications, and the total number of organisms that are replicating. What Behe and Snoke realized is that you need at least two point mutation changes in the gene code to get the most minimal change in a protein structure, and often you need three or more. Thus while you get one of your needed amino acid changes you have to wait for further replication cycles to get another point mutation change of just the "right" type, and at just the "right" position. In the meantime, while waiting for the second point mutation you hope that random chance doesn't "undo" the first point mutation before you get the second needed point mutation, and get a protein feature that natural selection can select or "fix" in the population. It is like comparing the odds of getting 6 while rolling one die (1 in 6 odds) as compared to getting two sixes while rolling two dice, (1 in 36 odds), except that the odds are much, much worst. What are some of these protein features that natural selection could possibly select for and "fix" in a population of organisms? One of the simplest is a disulfide bridge that holds two different points of a protein close together.

17. Verbal presentation by David W. Snoke, 60th annual Meeting of the American Scientific Affiliation at Messiah College, Grantham, Pennsylvania, August, 2005.

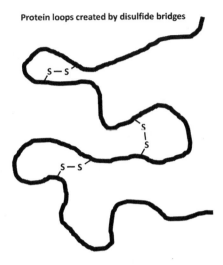

Protein loops created by disulfide bridges

Cysteine is the one amino acid that has as its side chain or appendage a –SH group. In chemistry we call this a thiol group. If you have two cysteine amino acids positioned just right along a protein chain and spaced far enough apart, then you can bend the protein strand into a loop, and with the right enzymes and helper proteins bind their two SH groups into a single disulfide bridge after the hydrogens have been removed.

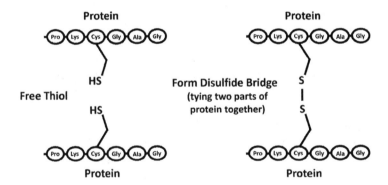

This creates loops and other secondary structures in a protein, which helps turn that particular protein from a non-functional (useless) protein into a functional one. But you have to get both cysteine amino acids into both positions at approximately the same time while you are mutating that protein chain. This drops the odds of getting the two amino acids together in the right place way, way down from just keeping one amino acid in the right place. Behe and Snoke called this a multi-residue mutation (residue mean amino acid in a protein strand). They did the modelling and found out that in general if you had an animal that reproduces once a year that had a population size of 1 billion, you should expect 50 percent of the animals to have evolved a new disulfide bond in a couple of proteins every 100 million years. If this sounds like an incredibly, incredibly slow rate for evolutionary change to occur, you are underestimating the severity of the situation. Between humans and apes, the differences between the proteins in our bodies are easily in the hundreds of thousands. According to this mathematical model how could random stochastic evolution account for these differences?

What kind of response did they get with their published paper? *Protein Science* has its origins in James Watson's research (of Watson and Crick DNA fame) back in the early 1970s. Since the journal's inception 2 protests have been filed against two different articles published in *Protein Science*. Behe and Snoke's paper drew over 1000 protests. This is unheard of. I, the author, cannot remember a protest filed against any physical chemistry oriented journal paper over my entire working career. What is it about this simple evolutionary modeling study that drew such overboard angst and ire? Are there some models scientists are not allowed to question?

Mathematical models are just models and only as good as the assumptions used to generate them. But such a reaction is bound to generate curiosity. Behe investigated further to see what experimental evidence would support the general conclusions of the Behe and Snoke modeling study. The result was his new book, *The Edge of Evolution—The Search for the Limits of Darwinism*[18].

Behe starts out stating evolution depends on three key factors—natural selection, common descent, and random mutation. In the modern scientific mind all three are closely linked and responsible for producing the diversity of life on earth. In the past the three were not so linked and the idea that animals shared a common ancestor was not coupled to the fact that random mutation with natural selection produced these connected animals. In short common descent, natural selection, and random mutation were three completely separate concepts that stood on their own over a limited working area of biology. In the modern era they were merged to produce the theory of everything, the theory of life. But it was not until the 1990s or later that science possessed the tools to evaluate whether all three working

---

18. Behe, The Edge of Evolution.

together can really do this, i.e. does the edge of evolution include the ability to generate unique new types of life?

When life reproduces or replicates, DNA is copied, but generally the copying of DNA is extremely faithful. On average a mistake is made only once for every hundred million or so base pairs (or nucleotides) of DNA copied in a generation. There are some exceptions where the mutation rate is sped up, such as in HIV and in anti-body production to meet foreign viruses. It has been suggested by some researchers that the increased speed was programmed in and constrained to a very small genome space to make the process successful. But for the generation of new life, the average one in a few hundred million copying error rate is accepted as realistic. Thus to realistically test whether random mutation can produce the miracles of life we need millions of replications, hundreds of thousands or more generations of an organism, and a huge population base. We also need almost complete knowledge of just exactly what was evolved and *do this at the molecular level*. Thus until recently it has been impossible to rigorously test the modern theory of evolution. But as Behe argues, now we can.

> One difficulty of writing a book questioning the sufficiency of Darwin's theory is that some people mistakenly conclude you're rejecting it in *toto*. It is time to get beyond either or thinking. Random mutation is a completely adequately explanation for some features of life, but not for others. This book looks for the line between the random and the nonrandom that defines the edge of evolution.[19]

This seems like a reasonable and perfectly scientific question to ask. Then why was Behe not considered a good choice for NSF or NIH funds to research this topic? I will argue that the key problem is the idea that nonrandom evolution may exist. The opposite of random is nonrandom or "designed," and that implies a designer, and that is not allowed in our current scientific environment—because we have introduced teleology into science, (even if it might be real). The materialist rightly understands that placing a limit on random mutation; also says that there is nonrandom or designed evolution. Once you introduce an outside force, intelligent agency, God into any aspect of evolution, it ceases to be evolution per say. This is because the move from God changing the right gene nucleotide base in a very slow process to God creating new species and families in progressive stages over long periods of time to God creating everything in 6 days is not that much of a philosophical or scientific stretch. Once you let the divine foot in the door, how can you stop it from coming in all the way? And with this higher

19. Behe, *Edge of Evolution*, 14.

agency involved it is hard for science to make definite statements as to exactly "*how*" this higher agency did it, unless we can "*scientifically explore*" this higher agency. To the theist, the scientific exploration of God and his works may be a "limited" but still "valid" line of inquiry. To the materialist it is utter nonsense. Behe has drawn a line in the sand with huge stakes for everyone of all philosophical persuasions. Is there a point where you can argue random evolution will not work? At that point the materialist has no explanation for how life developed and appeared.

Where Behe goes next is to search for an experimental test case to support and illustrate the model calculations of Behe and Snoke. He found a classic one in the parasitic disease malaria. Malaria is often called the million-murdering disease because it kills over a million people a year, (mostly children). The malaria cycle starts out as the single cell parasite *Plasmodium falciparum* (most virulent species of malaria). A mosquito infected with this parasite bites a human and injects into the human's blood stream the sporozoite phase of the malaria parasite cell. There it goes about many transformations and stages that will be greatly simplified here. Once in the bloodstream it makes it way to the liver where it multiplies some, and then continues on in the bloodstream. The malaria parasite cell then grabs the surface of a red blood cell, binds tightly to it, pulls itself inside the red blood cell and then wraps itself in a protective protein coat. Next it feeds on the hemoglobin; the cell can become stuck in the veins and stop circulating, the malaria inside the dying red blood cell reproduces until twenty or so copies of the malaria parasite cell are made, which burst out of the cell to infect new red blood cells. Another mosquito bite transfers the parasite to another human host as the old human host slowly dies of malarial complications.

What makes malaria a great random evolution test case is that this disease has been around in humans for thousands of years; infecting over a billion people per year, with over a trillion malaria parasites per person. We know the molecular basis for the mutations the human genome has come up with to slow or stop this deadly disease. And we know to the very amino acids in which protein; what molecular changes the malaria parasite *Plasmodium falciparum* has evolved over the past 60 years to become resistant to anti-malaria drugs. This allows Behe to put a mathematical limit on the number (generations) of the malaria parasite it takes to generate a beneficial random mutation that allowed it to resist our anti-malarial drugs. With that you have a generational and population size rate of beneficial mutations that can be extrapolated to other living organisms.

The anti-malarial drug atovaquone required the malaria parasite to mutate only one amino acid at position number 268 on one protein to give it resistance to this drug. Only 1 trillion malaria parasites or 1 in 10^{12}

generations were needed before the necessary random mutation occurred. It took only a decade or less before resistant strains of malaria developed against this drug. The malarial parasite *Plasmodium falciparum* has about one hundred million base pairs or nucleotides in its genome, so you get about one copying error per generation. However the famous malaria drug chloroquine required about 40 years—from 1940 to 1980 before malaria developed resistant strains. Doing the math, that equals about a hundred billion billion generations or 10^{20} generations of the malaria parasite. We now know that this malarial resistance to chloroquine was conferred by the mutation of two amino acids at positions 76 and 220 of the PfCRT protein in the vacuole (stomach) part of the *Plasmodium falciparum*[20]. What do these numbers mean? It took 10^{20} malarial cells to come up with one beneficial two amino acid substitutions. But if we buy human evolution and go back to the first primate in the human line 10 million years ago we can estimate that only one trillion or 10^{12} primates and humans have ever existed. Thus only *one* beneficial two amino acid substitution evolved us from earliest primate to *homo sapien*? That is all you can expect of random mutation in 10^{12} generations. Primate to *homo sapien* with one beneficial type two amino acid mutation. Really? Obviously something is very wrong with the creative powers of random evolution. Let us reproduce the data from Behe's Table 7.1[21] from the *Edge of Evolution* in the modified table below.

Approximate Size of Population and the Likelihoods for Some Events

Size of Population—Total number of organisms

Primates in the line leading to modern humans—past ten million years	10^{12}
Malaria cells in one sick person	10^{12}
Malaria cells worldwide in one year	10^{20}
Bacterial cells in the history of life on earth	10^{40}

Possibility of Event by Random Mutation—Average number needed:

Malaria cells needed (reproduction events) to generate atovaquone resistance	10^{12}
Malaria cells needed (reproduction events) to generate chloroquine resistance	10^{20}
Estimated organisms needed to generate one protein-binding site	10^{20}
Estimated organisms needed to generate two protein-binding sites	10^{40}

20. Bray, "Defining the role of PfCRT", 323–333.
21. Behe, The Edge of Evolution, 142.

Number of Random Mutation generated Protein-Binding Sites

Humans (10^8 humans needed)	1
E. coli (10^{13} E. coli cells needed)	0
HIV (10^{20} viruses needed)	0
Malaria (10^{20} organisms needed)	0

Typical Cell—Number of Protein Binding Sites Found	10,000

A protein-binding site is where two proteins are bound to each other, either by disulfide bridges connecting them or two amino acids one on each protein that hydrogen bonds to the correct amino acid hydrogen bond receptor on the other protein. At minimum you need at least two mutational changes in the two amino acids which are located on each protein to get a minimal binding site. Behe discusses in his book the experiment with breeding 10^{13} E. Coli cells in Professor Richard Lenski's lab[22] at Michigan State University. The last (and intermediate) generations of E. Coli had their genome sequenced to see what changes had occurred. The last bottom section of Behe's Table 7.1 and our modified version here summarizes the *experimental* knowledge we have of random evolution to generate new protein-binding sites using very large population sizes and many generations—while analyzing the changes made at the molecular level. The results are fascinating:

Malaria—over the course of human existence malaria has never been able to mutate in the necessary changes to overcome the sickle cell trait of African humans that makes them immune to malaria. No new protein binding sites have developed.

HIV—over the course of that disease in humans many mutations have been wrought making it immune to our drugs, but no new protein binding sites that confer new innovative properties have evolved[23].

22. Behe, The Edge of Evolution, 141.

23. In later arguments and comments posted at various places on the internet, Behe conceded that he may have missed one possible protein bridge that developed by random mutation in the HIV virus. A quick examination of the numbers and magnitudes in the Table 7.1 data, particularly 10^{20} HIV viruses, make 1 protein bridge evolved very consistent with Behe's calculations and results. Thus to keep the table data consistent with the table in The Edge of Evolution the protein bridge number for the HIV virus was kept at zero. Putting a one for a zero there makes no difference in Behe's conclusions.

E. Coli—over years of breeding in very fast generational turnover times (~20 minutes), mutations have been observed that that break systems (lose ability to make ribose—saves energy in well fed environments, etc.), but no new protein binding sites have been observed.

Humans—In our battle with malaria changing one amino acid at the sixth position of the second chain of hemoglobin produced sickle cell anemia. Deadly to children that inherited it from both parents, but life-saving to those that only got it from one parent. It stopped the malaria parasite. Behe only credits humans with one potential protein–binding site produced by random mutation over our existence. Interestingly, malaria has never found a way to beat the sickle cell trait.

Most proteins in the cell work as teams of a half-dozen or more, and that requires dozens and dozens of protein-bindings sites that have to be evolved by random mutation. How is random mutation going to pull this off with as few generations that we have with most organisms on the earth, particularly mammals, primates and humans? Behe sums it up with this quote.

> If the great majority of cellular protein-protein interactions are beyond the edge of evolution, it is reasonable to view the entire cell itself as a nonrandom, integrated whole—like a well-planned factory, as National Academy of Sciences president Bruce Alberts suggested. This conclusion isn't a rare property of just a handful of extra-complex features of the cell. Rather, it encompasses the cellular foundation of life as a whole.[24]

Behe is arguing that almost everything in the cell is the result of non-random evolution. The cell was basically "designed."

Lastly, we finish Behe's work with his earliest and perhaps most famous concept; the amazing complexity of the cell and irreducible complexity. In 1996 Behe published *Darwin's Black Box—The Biochemical Challenge to Evolution*[25]. The book went on to be a *New York Times* Bestseller, cited in *National Review* as one of the best 100 non-fiction books of the century, and *Christianity Today* gave it a book of the year award for the year 1997. The title of the book sets the stage. Until the microscope was invented the biological cell has been a "black box." But even with the invention of the optical microscope, the cell was still a black box at the biochemical level. It took the invention of the electron microscope in the 1930s to see down to the macro-protein level, and the invention of x-ray crystallography in the same era to elucidate the actual chemical structure of proteins, DNA, RNA,

24. Behe, The Edge of Evolution, 147.
25. Behe, Darwin's Black Box.

and other biomolecules. At that point Behe argues we have penetrated the last "black box" and have gone all the way down from biology, to molecular biology, to biochemistry, to chemistry. Thus if evolution is going to work it had better work on the molecular level or there is little point in discussing it as a viable theory. It was when we opened the last biochemical black box of the cell that we glimpsed a world of bewildering complexity and nano-sized molecular machines that defy Darwinian evolutionary explanations.

For macroevolution to be a valid model or theory of the origin of life it must explain the creation of life and provide a detailed mechanistic chemical explanation of how it happened. These sets of chemical mechanisms must allow us to make predictions about exactly how and where this evolutionary process could and could not occur, i.e., show me specific chemical reactions, and scenarios that work or forget the "last box," and the whole house starts to crumble. Behe argued that for a system to be built by a random mutation driven evolution then you must be able to provide a gradual path for a complex organism to be built. Each stage of this gradual path must be a viable working biological system with a functional advantage for which natural selection can optimize. In other words, it must be cumulatively complex. A system *is cumulatively complex* if the components of the system can be arranged sequentially so that the successive removal of components never leads to complete loss of function. Such a system always has some functional advantage to be selected for, no matter how slight. *Irreducible complexity* can be explained by contrasting with its opposite, *cumulative complexity*. In an irreducibly complex system any sequential loss of the components leads to a total loss of function. Thus how can natural selection select for any of the necessary intermediate stages to build the complex system sequentially? A quick diagram may explain this a bit better. Below is a ten part system that is cumulatively complex. All of the parts work together to make the system—but each part contributes something to the functional advantage

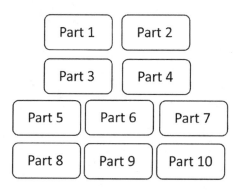

the system provides the organism, so that natural selection can select for it on successive generations of the organism.

Now we remove part 6. The whole system is hurt, but it still provides

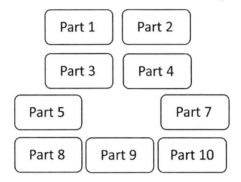

80 percent of the original functional advantage. There is still some functional advantage for natural selection to select.

Now we remove parts 6, 7, and 1. The system is badly crippled at this

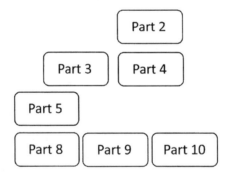

point, but still retains 35 percent of the functional advantage. Still something that natural selection can select.

Now all the parts are removed but part 9. The system hardly works, but there is still 3 percent of the functional advantage left in part 9. It is not much but there is still something for natural selection to optimize on, until

Part 9

the whole system can be built back up gradually in reverse order. This is a *cumulative complex* system.

But suppose this 10 part system was irreducibly complex. Every part is required before you get anything other than 0 percent functional advantage. If I remove part 6 or any part, I get nothing, no functional advantage. How can gradual random mutation driven evolution build such a system from its individual parts? There is nothing for natural selection to operate on during the construction phase of the system. This is a key point that is very frequently overlooked in criticism of Behe's ideas. Below is just such a case from a very vocal critic of intelligent design, P. Z. Myers.

> Complexity, complexity, complexity complexity. Oh look, there's a pathway—it's very complicated. Complexity! Complexity, complexity complexity—complexity. And did you know that cells are really, really complicated? But we're not done—complexity! Complexity (complexity complexity). And you're gonna be blown away by the bacterial flagellum—it's like a little machine! And it's really, really complicated! Complexity—complexity complexity. Complexity. We need more cells, they're really complicated. You just get blown away by these things, they are just so amazingly complicated. Complexity. Therefore; design.[26]

Later P. Z. Myers does try to address the irreducible nature of the problem but is more muted at that point. To get a better grasp of the irreducibly complex idea let us apply this to a known system, one with a known

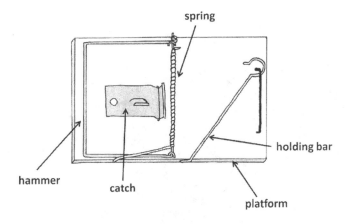

26. Myer, "Irreducible Complexity."

functional advantage or purpose, the famous irreducibly complex mouse-trap illustrated in Darwin's Black Box[27] and redrawn here.

The five basic parts of the trap are: the hammer, spring, catch, holding bar, and platform. Behe argues you must have these five functions (not necessarily parts, though the parts serve function), or the mousetrap just will not work. To get the functional advantage of catching mice, you must have all five parts working together or it fails. Many people have tried modifying the mousetrap in an evolutionary manner to show how it might have evolved gradually, but they fail to resolve the necessity of having all five functions working before the mousetrap does anything. For example, get rid of the platform; simply attach the parts to the floor. At that point you are using the floor as a platform, as a system to position the parts correctly so that they function together. Others have suggested doing away with the catch and sliding the end of the holding bar (once it has been placed over the compressed hammer) under a stray extruding part of the spring. But at that point you are using the spring as a "catch" in addition to its duties as a spring. Thus you must have the five functions to get the operation of a mousetrap. He then argues that if you see a biochemical system that is similarly irreducibly complex like the mousetrap is, then you have identified an irreducibly complex system that begs Darwinian explanation. Do such irreducibly complex systems exist in the cell and other biological systems? Behe argues yes, there are many such systems and spends much of the book going over them in detail. The best way to grasp the enormity of the irreducible complexity present

27. Behe, Darwin's Black Box, 43.

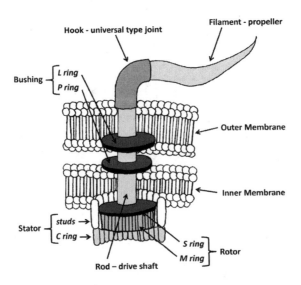

in the cell is to read *Darwin's Black Box*. *The Edge of Evolution* updates and introduces even more mind boggling systems. Here I introduce two of Behe's examples to get the point across in the biochemical domain.

The bacterial flagellum as shown in the diagram[28] is essentially an outboard motor that propels the bacteria from one location to another in the aqueous environment of the bacteria. It is housed in the outer membrane of the bacterial cell. Not every cell or bacteria has flagellum, only a subset species of bacteria do. But for those that do the cellular motor properties are amazing. These biochemical motors spin at 17,000 RPM or 283 revolutions per sec. At full speed they can stop their direction of spin and spin in the opposite direction in less than one revolution. It does not use electricity or gasoline but instead a flow of biomolecules from the cell fuels the spin chemically and mechanically in a rotary chemical process that imparts motion to the turning commutator part of the motor. But the comparison to a human built electrical motor is uncanny. Human electrical motors require a motor casing or shell to keep the various components oriented and stable for proper function. Likewise the flagellum motor has the two bi-lipid membranes (*inner and outer membranes*) to serve as the motor casing. Electrical motors require a turning shaft to conduct rotary power and two rotary bearings housed in the motor casing to hold the turning shaft. Likewise the flagellum motor has two bearings or bushings (*L ring* and the *P ring*) that hold the rotating shaft in place as it turns. The rotating shaft in an electrical motor has a commutator set of magnets or windings built into it to receive electrical force from the motor casing bound or stationary wire windings. The flagellum motor rotary shaft has a set of *S and M rings* to receive chemically induced rotatory force from the stationary and motor casing bound *studs* and *C ring*. Outside the cell attached to the rotary shaft is a flexible protein joint (the *Hook—universal joint*) that connects a long *filament* to the rotary motion. Upon spinning it acts as a propeller that moves the bacteria quickly in the opposite direction of the rotary shaft and propeller. It is like an electrical motor connected to a propeller in a trolling motor that moves the boat (bacterial cell) in the opposite direction of the water being ejected by the propeller.

The electrical motor must have the basic parts assembled and in position before it can do its function—rotary motion. These are the motor casing, the bearings, the rotary shaft, the commutator, and the stator. Without all of these basic parts in proper orientation the electrical motor will not work. Likewise the flagellum motor requires a casing, bearings, rotary shaft, commutator and stator, or it will not work. Thus the parts alone do not have

28. Behe, *Darwin's Black Box*, 71.

the functional advantage of any degree of rotary motion, but instead impart a negative drain on the resources of the bacterial cell to produce them. Thus no functional advantage is imparted and natural selection should eliminate them before the flagellum motor is ever constructed. The flagellum motor needs to be evolutionarily evolved as a *complete unit* before any functional advantage of bacterial motion is obtained. How does stochastic gradualistic evolution accomplish that?

It is fascinating to note that almost all competent criticism of Behe's flagellum example focuses on possible ways the individual parts may have some other functional advantage along the way or are co-opted from other working biochemical systems. No one disputes the fact that gradual random evolution is incapable of assembling the flagellum motor all at once. Thus if the flagellum motor really is irreducibly complex we have a show-stopper on our hands.

The criticism of Behe's flagellum motor falls into two basic arguments. One, finding some other kind of functional advantage for the smaller parts, which often is not experimentally demonstrated, but given as *hypothetical* scenarios. The other option given is the co-option argument[29],[30]. If a similar biochemical system such as the Type III secretion system (TTSS) has many of the protein structures used by the flagellum motor, then it can be *co-opted* by evolution to construct a flagellum motor. The TTSS is a needle-like structure that pathogenic germs such as Salmonella use to inject toxins into living cells. The base of the needle has ten elements in common with the flagellum. However, forty of the proteins that make a flagellum work are missing from the TTSS. What is not explained is how you can convert a linear sliding motion of two or three inter-fitting tubes into the rotary motion of just one rod. How would you do that in a step-by-step Darwinian process that preserves some kind of functional advantage, superior to the previous functional advantages of the starting system? It is like saying a motorcycle has many of the same parts as a car, thus random evolution can co-opt motorcycle parts to build a car. Maybe you can in an irreducible machine shop with appropriate intelligent personnel. But where do they come from? A similar situation exists on the biochemical level in transforming a TTSS into a flagellum.

The next time you cut yourself stop and stare at your wound. Something amazing is going to happen over the next five minutes. You are not going to die. Why? The blood clotting cascade of mammals is going to save your life. See the diagram adapted from *Darwin's Black Box*[31].

---

29. Myer, "Irreducible Complexity."
30. Miller, "The Collapse of Intelligent Design."
31. Behe, Darwin's Black Box, 82.

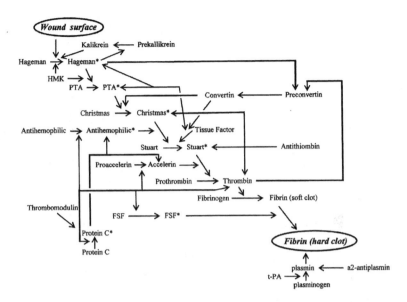

At the wound surface a protein called *Hageman* protein interacts with the *Kallikren* and *Prekallikren* proteins in a small feedback loop to activate the *Hageman* protein to the *Hageman** protein which interacts with the *PTA* protein to the *PTA** protein while setting up another larger feedback loop with *Preconvertin* to *Convertin* proteins which interact with the *PTA** to *Christmas* proteins to set up the *Christmas** protein on down the protein cascade of *Christmas** to *Stuart* to *Stuart** to *Prothrombin* to *Thrombin* to *Fibrinogen* to *Fibrin* (soft clot) to *Fibrin* (hard clot). In these series of sequential protein reactions we have skipped at least six other feedback loops involving other proteins. Also ignored was the crucial interaction of the *Antihemophilic* to *Antihemophilic** with the *Christmas** protein to activate the *Stuart* to *Stuart** protein part of the blood clotting cascade. All of this is occurring in your bloodstream at the wound site, and occurring fairly quickly. If it did not you would bleed to death. Then another amazing thing happens. Once the wound is plugged up with a blood clot the clotting stops. If it did not, clotting would continue until an important artery or vein was plugged or a piece of the overgrown blood clot would break off, enter the heart or brain, and kill you with a heart attack or brain stroke.

Up until a few years ago, humans who were hemophilics died an early death; all of them. Most hemophiliacs do not produce enough *Antihemophilic* protein, and most of the rest do not produce enough *Christmas* protein. We give hemophiliacs a few more decades of life with massive infusions of these proteins every time they bruise or cut themselves. Without modern medicine these poor souls would be dead. If insufficient quantities of just one of these two proteins will stop the blood clotting cascade, then what about the rest? Have we identified another very irreducible biochemical system?

Arguments against the irreducible blood clotting cascade seems to center on finding other mammalian blood clotting cascades that differ by one protein, or in one fish system, four proteins, and then appealing to evolutionary computer program simulations that appear to breech this problem[32,33]. But the dissenting responses are even more significantly muted than the ones offered against the flagellum biochemical system. Reading dissenting responses to irreducible complexity and Behe's example systems can be very entertaining. Steve Mirsky in the February 2005 issue of Scientific American says,

> Their co-conspirators, the "intelligent design" crowd, go with the full-blown intellectual surrender strategy-they say that life on earth is so complex that the only way to explain it is through the intercession of an intelligent super-being[34]

Co-conspirators and full-blown intellectual surrender strategy? Really? The question is: why such anger and ire in a professional journal over what ought to be a professional discussion of the merits of random mutation to build complex biochemical systems? My opinion—a worldview was severely challenged, and you got a religious reaction rather than a measured academic one. Here is quote from the cover jacket of the first edition of *Darwin's Black Box* that sums up the present biochemical situation.

> For Darwinian evolution to be true, there must have been a series of mutations, each of which produced its own working machine, that led to the complexity we can now see. The more complex and interdependent each machine's parts are shown to be the harder it is to envision Darwin's gradualistic path. *Behe surveys the professional science literature and shows that it is completely silent on the subject, stymied by the elegance of the foundation of life* (emphasis added)[35]

32. Myer, "Irreducible Complexity."
33. Holmes, "Creationism special."
34. Mirsky, "In the beginning was the cautionary advisory."
35. Behe, Darwin's Black Box, cover jacket.

9

Cambrian Explosion and the Tree of Life

It was dark in the small auditorium and there was a small crowd. The speaker was a well-known American professor from a Christian university who had written a favorable history of the intelligent design movement. The small crowd was Hungarian; we were in Budapest, Hungary. The Hungarian Lutheran Seminary professor, who brought me to this lecture, was Hungarian. He spoke excellent English in addition to his native Hungarian and other languages. God, I envied those who speak two or more languages fluently. There is joke that says a tri-linguist is one who speaks three languages, a bi-linguist is one who speaks two languages, and one who speaks one language is called an American. It was time to start the lecture.

The main guest speaker had brought another American along, a medical doctor interested in creation/origins issues. He started first as the warm-up speaker. The good doctor started with lines to the effect that the age of the earth was not as old as secular education has taught us, rates of salinity in the ocean, problems with the fossil record sequence; interesting but standard young earth creationism lines that I have heard before. He finished with a flourish and then turned the podium over to the main speaker.

The main speaker being a communications professor was a bit more at home in front of the crowd. He started with a brief introduction and then said, "I really need to apologize to my colleague. What I am going to say is going to contradict some of his talk and he had no idea I was going to do this. I used to be a young earth creationist; but I have decided to go public today and switch to an old earth creation perspective where God used 14 billion years to create the universe, but then found an appropriate solar system with a suitable planet and swooped in on that planet and created our world as we know it in six literal 24 hour days as described in Genesis." He moved on to give a well prepared discussion of the finely-tuned parameters physicists have discovered for our universe; and how they point to design and evidence for God.

His doctor colleague looked like a deer caught at midnight on the super highway in the glare of five transfer truck headlights. I would not have traded places with him for a thousand dollars. The main speaker seemed like such a nice guy, how could he do this to his colleague? I did not know whether to roll on the floor in laughter or cry in sympathy.

After the talk ended, (I have no idea how the Hungarians responded to this origins perspective), I rushed over to wait in line, and ask him a few questions. I did not see his doctor colleague friend; I guess they were taking him out on a stretcher, or else he was looking for a gun or sharp knife, who knows. When the professor was free I cornered him and said, "That is one of the most unusual origins perspectives I have ever heard. Where did you get it from, why did you change?

"Oh I got it from a book. I am committed to a 6 literal day creation, but I am so fascinated by all of the fine tuning of the universe that the intelligent design movement is using to build a case for design. So this allows me to do both."

I asked, "How do the young earth creationists and old earth creationists react to you?"

"Oh they both hate me."

Our conversation continued a while further, and he gave me free of charge a book and other materials he had written, that I gave to my Lutheran professor friend. My Lutheran friend was bowled over. He was desperately trying to find information on the American intelligent design movement and not having much luck. He was writing his habilitation thesis for his Doctor of Science degree on creation studies, and intelligent design had just burst on the international scene. In some European countries (like Hungary), you can get an additional advanced degree beyond the Ph.D. called the Doctor of Science degree. It had been a very interesting evening, and I went home to my apartment on Bela Bartok Ute that night, to ponder this interesting turn of events.

In 1995 Time magazine ran a cover story on fascinating new fossils found in the Cambrian-era Moatianshan shale outcrop in Chengjiang County, Yunnan Province of the People's Republic of China. Some scientists suggested that it turned Darwin's theory of the gradual development of planet and animal life upside down, quite literally. It turns out that Darwin's grand scheme of evolving life was incongruent and problematic with known fossil finds of the Cambrian era even in Darwin's day; even when he wrote *Origin of Species*. Stephen Meyer tells the fascinating historical and scientific tale in *Darwin's Doubt*[1].

In *Origin of Species* Darwin suggested that 1—randomly arising variations, 2—the heritability of those variations, and 3—a competition for survival, resulting in differences in reproductive success among competing organisms[2] produced the fossil record and tree of life. As discussed earlier, this was modified in the 1920–30s into the neo-Darwinian Synthesis. But either Darwin's version or the modified neo-Darwin version of gradually accumulating change would produce new species first. This would be followed by groupings of new species into genus, then genus into families, families into orders, and then the production of phyla from the orders. Many similar species represents *diversity*. The different phyla with their huge body plan differences represent the *disparity* between the phyla and orders. According to the modern theory of evolution *diversity* should precede *disparity* and the fossil record should record this fact. It is also understood that very long periods of time, millions and millions of years, are needed before evolutionary *diversity* produces the depth of *disparity* that we now see. Meyer includes a diagram[3] in *Darwin's Doubt* that illustrates this nicely. Shown below is a modified version of the diagrams previously referenced in *Darwin's Doubt*. The top graph shows the classic tree of life expected by Darwinian theory, where one or more species diversifies with time (vertical axis), into more and more morphology (horizontal axis). The lower graph below it split into the Precambrian, Paleozoic, Mesozoic, and Cenozoic era shows time on the vertical axis, and likewise morphology or *disparity* on the horizontal axis. As you compare the two graphs it becomes obvious from a phyla viewpoint that most phyla of life started at almost the same time (the Cambrian period).

None of the distinct branching pattern expected is shown in the fossil record. It is as if most of the basic body plans of life, i.e. the phyla, appear on the scene at once in a sudden burst of creativity, and then they continued on from there to diversify. It is like *disparity* came first followed by subsequent *diversity*. It is upside down from the expected Darwinian pattern. Top down instead of bottom up.

1. Meyer, Darwin's Doubt.
2. Meyer, Darwin's Doubt, 10.
3. Meyer, Darwin's Doubt, 10, 35, Figure 2.7.

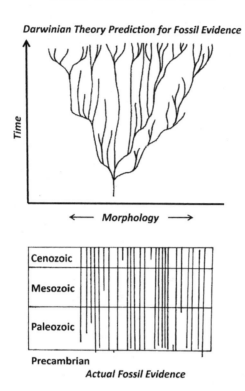

Darwinian Theory Prediction for Fossil Evidence

Darwin was well aware of this problem. In 1859 knowledge of Cambrian fossils was limited, but it was known that few if any fossils predated the Siluran period (name used for the Cambrian period in Darwin's day). So the sudden appearance of life in the fossil record was becoming apparent, even in Darwin's day. He wrote in *Origin of Species,*

> The difficulty of understanding the absence of vast piles of fossiliferous strata, which on my theory were no doubt somewhere accumulated before the Silurian [i.e. Cambrian—modern name] epoch, is very great . . . I allude to the manner in which numbers of species of the same group suddenly appear in the lowest known fossiliferous rocks.[4]

Darwin sent a copy of *Origin of Species* to Harvard paleontologist Louis Agassiz, arguably one of the best scientists of his day, who knew the fossil record better than any man alive. Agassiz is famous for the story of the graduate student he made observe a preserved dead fish until he could see hundreds of details and features. Agassiz was a trained empirical observer

---

4. Darwin, *Origin of Species,* 306–7.

with few if any equal. Instead of gaining an ally to his theory, Darwin gained a nemesis. Agassiz became Darwin's nemesis and refused to endorse Darwin's theory, because of its contradiction with the Cambrian (Silurian) fossil record. Meyer described Professor Agassiz as a German idealist and quotes historian A. Hunter Dupree, "'Agassiz's idealism was of course the basis of his concepts of species and their distribution,' of his insistence that a divine or intellectual cause must stand behind the origin of each type.[5]" "Intellectual cause behind the origin of each type," a fascinating idea to come from the dean of paleontology in 1872. It almost sounds like something from the modern intelligent design movement. In 1872 anyone claiming that Agassiz was not espousing science, would have been laughed at. Today it has become almost the reverse. Does our definition of science change that easily?

In Darwin's time Wales, England was the source of Silurian or Cambrian fossils. Darwin even explored them in his student days with the Cambrian fossil expert Adam Sedgwick. Darwin was criticized by heavily Agassiz and Sedgwick on how his theory could not accommodate the abrupt appearance of life in the Cambrian period. To his credit Darwin acknowledged this criticism in *Origin of Species,*

> To the question why we do not find rich fossiliferous deposits belonging to these assumed earliest periods prior to the Cambrian system, I can give no satisfactory answer . . . The case at present must remain inexplicable: and may be truly urged as a valid argument against the views here entertained.[6]

But Darwin did not let these criticisms stop his theory, his answer, *the fossil record is incomplete*; and time and new fossil discoveries would validate his theory. In fact the fossil record did validate Darwin to some extent, more fossils were found and some transitional looking fossil specimens were discovered.

However the general trend of the fossil record still remained the same. New species and families of fossils come on the scene, stay basically the same for millions of years, and then disappear. There is little hint of the millions of transitional species suggested by Darwin's theory in the fossil record. A few are there but largely they are very absent, particularly among the animals. At the Precambrian—Cambrian boundary the fossil problem became acute. So acute that it was nicknamed the Cambrian Explosion. Explosion because the geologic time span that generated over 20 of the 26 phyla of life was shortened to a time span of 20, 10, possibly 6 million years. Life

5. Dupree, Asa Gray, 227.

6. Darwin, Origin of Species, 308.

on earth expanded from 3 phyla to 20 phyla in less than 20 million years. It was an explosion of creative life development that boggles the mind. It creates enormous problems for Darwin's theory of gradual development. Figure 1.8[7] from *Darwin's Doubt* illustrates the problem. A recast version is shown here.

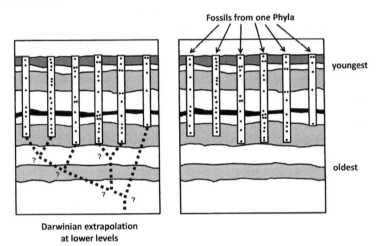

On the right is the fossil layer boundary diagram of the Precambrian/ Cambrian period, where time is the vertical axis with the oldest layers being at the bottom (Precambrian period), and the newer layers at the top half (Cambrian period). The vertical white bars represent different phyla of fossils, and the dots the actual fossils found. These diagrams are simplified but show that most of the phyla of life start in almost the same fossil layer. The diagram on the left is the same but modified to show what you would expect from Darwin's theory. Not all phyla start at the same time, and there should be connecting branches and nodes between the different phyla that are represented in the fossil record with some kind of distinct fossil. As noted by the questions marks in this diagram, they are not there. If one can justify the missing fossil's absence by sufficient reasons, (why these animals were not fossilized?), then the diagram on the left, and Darwin's theory can still withstand the Cambrian explosion problem; not as proved theory, but still a possible viable theory. If not, then Darwin's theory has a massive problem. Meyer devotes the rest of *Darwin's Doubt* to explaining why these missing fossils are really missing, i.e., not there; and what are the ramifications.

The tale of the Cambrian fossils is not complete without two major Cambrian fossil finds that defined the bizarre animal world of the Cambrian

7. Meyer, Darwin's Doubt, 24.

period. In 1909 the director of the Smithsonian Institute Charles D. Walcott discovered the Burgess Shale fossil field. The Burgess Shale refers to a fossil-rich area on Fossil Ridge between Wapta Mountain and Mount Field, just a few kilometers north of Field, British Columbia, Canada. Over the ensuing years Walcott's team excavated over 65,000 specimens of an astonishing level of preservation. Often soft body parts were clearly defined. Brachiopods and trilobites were known since Darwin's day, but Walcott added many more bizarre species. For example, three of these, *Marrella*, *Opabinia* and *Hallucigenia* were so bizarre that they defied reasonable classification for decades. For example the aptly named *Hallucigenia* is like a monster out of science fiction; consisting of a round mass at one end (head?) that had a cylinder shaped trunk with seven pairs of spines shooting upwards on either side. On the underside are seven pairs of limbs correlating to the spines on the top. At the underbelly before the trunk are three pairs of shorter tentacles that curve upward. Each appears to have a hollow tube connected to the gut and is outfitted with a pincer at the tip. No one knew where to classify this bizarre animal.

Walcott sided with Darwin, not Agassiz. So he attempted to find the precursors of the Burgess Shale menagerie, with little success. His answer to this puzzling riddle was to suggest that at low sea levels the Precambrian fossils were deposited, and then after the sea rose the Cambrian fossils were deposited in the Canadian Rocky Mountain layers. These layers were eventually uplifted while their Precambrian precursors remained buried deep in sediments under the ocean floor. The transgression and regression of the seas made the precursors of the Cambrian fossils inaccessible. Future exploration of the fossils below the ocean floor would make these fossils accessible; if and when technology made that possible. The Burgess Shale specimens went to the Smithsonian Institute files where many of them remained far from public view, until decades later.

The next Cambrian era find was more recent and just as exotic. Near Chengjiang County, Yuxi City, Yunnan Province of the People's Republic of China is a shale outcropping that bears Cambrian fossils older than the Burgess site and preserved even better. The fossils were first discovered and described by Henri Mansuy and Jaques Deprat in 1912. But it was not until 1984 that Professor Hou Xian-guang realized the significance of the find. The oldest layer dates to ~518 million years ago making it the oldest source of Cambrian fossils available. This fossil field contains all of the phyla found in the Burgess site plus new phyla. As of June 2006; 185 different fossil species have been identified, with 8 possible members of the phyla *Chordata*[8]. To grasp what this means; recall that all vertebrate animals, including fish

8. "Maotianshan Shales"

and humans, belong to the phyla *Chordata*. The phyla containing humans, the most complex creature on the planet, dates to the very dawn of life. How could this be? Where is the time to gradually evolve a *Chordata* from mollusks and worms? Obviously, understanding what fossils lie just below the Cambrian period, in the Ediacaran period becomes critical. If the precursors are not there, then the Cambrian explosion of life is very real, and Darwin's tree of life is utterly out of sync, as well as the theory.

Now is a good time to address Darwin's and Walcott's answer to: where are the precursor fossils to the Cambrian fossils? We continue with Stephen Meyer's arguments and references from his book *Darwin's Doubt*. Darwin argued the fossil record is incomplete. Since Darwin, 150 years of paleontological searching has greatly expanded our fossil knowledge. But the same incomplete structure, stasis or little change in groups as they appear and disappear; with little to no transitional fossils remains the pattern. Michael Foote a statistical paleontologist from the University of Chicago has published[9] that we have a representative sample of morphological diversity and therefore can rely on patterns documented in the fossil record. He argued, "in many respects our view of the history of biological diversity is mature[10]." Darwin's argument that the fossil record is incomplete is technically true, (we will find some new fossils), but statistically and meaningfully; *not true anymore*. We have found enough. We have a good picture of the fossil record. The discontinuous nature of the fossil record is *real*. What about Walcott's answer? The seas rising up and down did it. We now have the technology to drill for oil out in the ocean basin and discover the fossils down there. There are no Precambrian fossils to be found. In fact, plate tectonics recycles the ocean bed sediments such that Precambrian sediments will never exist out there. Walcott lived long before plate tectonics was discovered. If Precambrian sediments are to be found they will only exist on continents where we have found them previously (with almost no fossils). But the Cambrian/ Precambrian Moatianshan shale outcrop in Chengjiang, China does have an extraordinary collection of Precambrian (Ediacaran) fossils. And what it shows, or more importantly does not show, is stunning.

In 1998 Dr. J. Y. Chen found a fossil of a soft-bodied sponge in the Precambrian layers below the famous Moatianshan Shale. His U.S. co-researcher Dr. Paul Chien was able to examine the surrounding sponge fossil rock with a powerful scanning electron microscope to identify the tiny delicate sponge embryos that were also fossilized[11]. He could identify these

9. Foote, "Sampling, Taxonomic Description," 181–206.

10. Foote, "Sampling, Taxonomic Description," 181.

11. Chen, "Wen'an Biota."

structures down to the single cell level. The ramifications are profound. If the Precambrian layers can preserve soft-bodied animals like sponges and even their tiny eggs, then where are the precursors to all of the hard-shelled and skeletoned animals of the Cambrian period? Meyer describes it like this,

> Yet these Precambrian layers did not preserve remains of any clearly ancestral or intermediate forms leading to the other main groups of Cambrian animals. This raised an obvious question. If the Precambrian sedimentary strata beneath the Maotianshan Shale preserved the soft tissues of tiny, microscopic sponge embryos, why didn't they preserve the near ancestors of the *whole* animals that arose in the Cambrian, especially since some of those animals must have had at least some hard parts as a condition of their viability?[12]

The obvious answer; they did not exist or they would be there. Well are there any fossils in the Precambrian or Ediacaran period? Other than the sponge mentioned previously there are basically four types, a flat air mattress-like *Dickinsonia* fossil, *Spriggina* with body segments and a head(?) shield, a frond-like *Charnia*, *Kimberella* a very primitive mollusk and worm-like tracks that might have been made by *Kimberella*. Most of these fossils were found in the Ediacaran Hills of Australia. Meyer devotes chapter 4 of *Darwin's Doubt* to explaining why most paleontologists who are experts in this area do not see these fossils as anything that could be a precursors to the birth of animal life in the Cambrian period. Except for *Kimberella*, scientists are not even sure whether or not these fossils are animals or plants. Some may deserve their own separate phyla unrelated to any that exist today. Thus where is the precursor to *Chordata* and all the rest of the phyla that so suddenly appeared in the Cambrian period?

How abrupt is this Cambrian explosion of life? Here is a figure that was redrawn for clarity from Meyer and *Darwin's Doubt* that puts it into perspective.

Go to the bottom of the diagram where it says origin of earth 4600 million years ago (4600 MYA), or 4.6 billion years ago. If *Present* at the top is 0 million years ago, then divide the vertical time axis into 0 MYA, 1000 MYA, 2000 MYA, 3000 MYA, 4000 MYA, 4600 MYA. At the point 3500 MYA put a line and mark first bacteria. Now continue on up the graph and at the point of 570 MYA[13] put Precambrian (Ediacaran Period) fossils (about four different types found) appear. Now just above it at 543 to 490 MYA we put the

---

12. Meyer, Darwin's Doubt, 68.
13. Meyer, Darwin's Doubt, 73.

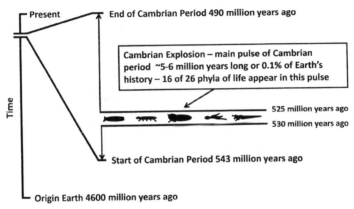

Cambrian period with its addition of 20 more phyla of life, and hundreds of new species of every imaginable type. Now within that 53 million year span of the Cambrian period we see several explosive spurts of new life. The most dramatic one is between 530 and 525 MYA or 5–6 million years long. During this time representatives of at least 16 completely novel phyla and about 30 classes first appeared in the rock record[14]. Let us picture it another way, using a stack of pennies. Each penny equals 5 million years.

Stack of 920 Pennies representing 4600 Million Years of the Earth's History

Stack of 98 pennies on top	End of Cambrian period until now
Third penny from bottom of previous 11 penny stack	Time for 62% of all of life's plans or phyla to come into existence
Stack of 11 pennies on top	Time for 20 phyla of life to appear; only 3–5 more are left to start later
Stack of 5 pennies on top	Time for the four animal/plant(?) Pre-cambrian fossils
Stack of 586 pennies on top	Time that only bacteria fossils are found
Initial stack of 220 pennies	Time since the earth formed until the first bacteria appears

Do you see Darwin's branching tree of life in this? Most people do not. How accurate is that one penny representing 5–6 million years when 62% of the basic body plans of life just appeared on earth? Meyer references

14. Meyer, Darwin's Doubt, 73.

Samuel Bowring, the MIT geochronologist, as saying 16 phyla appeared in 6 million years[15], and Douglas Erwin and colleagues showing 13 new phyla appearing in 6 million years[16]. Do not forget to add to the third single penny or 6 million year period at least 62% of all life plans appeared, the appearance of the first trilobite with its lens focusing compound eye. From *no* eye to a *lens focusing compound eye* in 6 million years; Charles Darwin would have thrown up his hands in utter frustration with such a timeline for the dreaded evolution of the eye.

How receptive is the paleontological community to these discoveries? It varies widely, depending on who you talk to and how they interpret what it means. But a fascinating story was given by Stephen Meyer[17]. The Discovery Institute of Seattle sponsored a talk by Professor J. Y. Chen (discussed earlier) on his Cambrian fossils finds at the Maotianshan Shales at the University of Washington Geology department. His fame and previous publications guaranteed considerable interest. Meyer continues,

> So there was little doubt about the significance of the discoveries that Chen came to report that day. However what was soon in doubt was Chen's scientific orthodoxy. In his presentation, he highlighted the apparent contradiction between the Chinese fossil evidence and Darwinian orthodoxy. As a result, one professor in the audience asked Chen, almost as if in warning, if he wasn't nervous about expressing his doubts about Darwinism so freely–especially given China's reputation for suppressing dissenting opinion. I remember Chen's wry smile as he answered. "In China," he said, "we can criticize Darwin, but not the government. In America, you can criticize the government, but not Darwin.[18]

Is there a way to explain the stasis and sudden appearance and disappearance of groups in the fossil record? Paleontologists Niles Eldredge and Stephen Jay Gould realized that absence of change, *stasis*, is data. They went on to propose a new theory of evolution of rapid changes followed by long periods of stasis called *punctuated equilibrium* (or Punk Eek! as it is affectionately called). They argued that natural selection operates more on competing species, particularly species separated by geographic or other features, than individual organisms within the specie. This was called allopatric speciation, and they argued it would produce much more dramatic

15. Bowring, "Calibrating Rates of early Cambrian Evolution," 1297.

16. Erwin, "The Cambrian Conundrum," 1091–97.

17. Meyer, Darwin's Doubt, 51–2.

18. Meyer, Darwin's Doubt, 51–2.

morphological changes, that traditional Neo-Darwinism. The recast figure below based on a similar figure[19] in *Darwin's Doubt* shows the difference Punctuated Equilibrium evolution would make on the tree of life as follows.

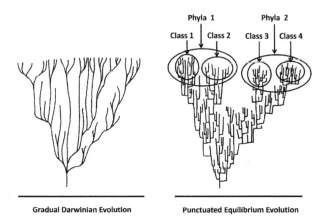

Gradual Darwinian Evolution   Punctuated Equilibrium Evolution

The tree on the left is the traditional gradually evolving tree of life. The tree on the right is what punctuated equilibrium would produce. The short horizontal lines are the rapid species producing events that leave few if any fossils. The vertical lines are the periods of stasis that leave relatively unchanging fossils, like what we see in the fossil record. As expected few transition type fossils would be found. But does this really solve the problem of the fossil record. Examine the next diagram inspired by a similar diagram[20] in *Darwin's Doubt*.

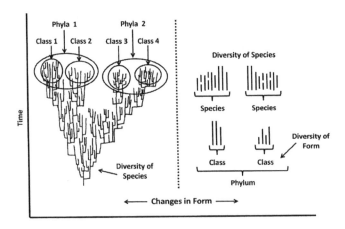

19. Meyer, *Darwin's Doubt*, 171.
20. Meyer, *Darwin's Doubt*, 171.

On the left is Punctuated Equilibrium evolution's tree of life. On the right is the general picture one gets from the fossil record, particularly at the bottom of the tree during the Cambrian explosion. Though Punctuated Equilibrium does provide a somewhat of an answer for the missing transition fossils, it still fails badly to explain the top-down, or phyla to specie aspect of the fossil record, particularly during the Cambrian and Precambrian periods, and to a lesser extent later on. Punctuated Equilibrium depends heavily on having many "species" from which natural selection chooses the "most fit," etc. And an abundance of species in the Precambrian period from which natural selection can 'select" for the Cambrian period is exactly what Punk Eek does not have. This was noted in a seminal paper by paleontologists D. Erwin and J. Valentine[21] in 1987, and little since then has changed.

The wind was a bit chilly as it swept around the brick buildings onto the long sidewalks. Ellie wrapped her sweater a bit tighter, comforted by the fact that summer was soon approaching. Tap, "Ellie."

"Aaaiii! Who?! What?!!"

"It's me, Bert!

"Gheez, you startled me!" Ellie had whirled around to see Bert and Mary walking together holding hands.

"I'm sorry; I did not mean to scare you." Bert apologized while Mary giggled.

"You and Jon are masters of that!"

"Hey where are you going?

"To meet Jon, in the Palace."

"Can Mary and I tag along?"

"Sure," Ellie really wanted some time alone with Jon, but that could wait for later. The threesome strolled among the buildings chatting lightly until they reached their destination, Som and Ela's Chicken Palace. Jon was there at a large booth, and seemed a bit surprised to see Bert and Mary.

"Come join us you two theist vagabonds!"

Bert smiled, Mary wryly, as they slid into the voluminous booth. "Did you read *Darwin's Doubt* by Meyer?" Bert wasted no time in asking the question.

"May I have your order, please?" the petite waitress was all business.

"Water and a small order of fries."

"Large coke and cheesy bacon fries."

"Just a small diet Coke for me"

21. Valentine, "Interpreting Great Development Experiments," 71–107.

"It's 11:00 am; I'm ready for lunch! One specialty order of chicken, and a Coke for me!"

"We don't serve chicken here," the waitress seemed genuinely perturbed.

"But isn't this the Chicken Palace?"

"Yes, we serve hamburgers and our specialty is pizza," the waitress was smiling broadly now.

So Jon flashed a smile back, "I guess I'll have a slice of your specialty pizza, and a medium Coke then." As the waitress left, Jon grinned at Ellie, and turned to Bert and Mary, "First you cram God down my throat, and now you want to take away my Darwin."

"No, no, I did not mean it in that way! Yes I would like to see you change your mind, but on your terms, not mine." Bert was alarmed at the snap from Jon and was unsure just how much he was joking or not joking. "Is there anything about it that bothers you that much?"

"Yes there is."

"And what is it?"

"The fossils, they are still there, and they do increase in complexity as you go up the tree of life, how does that fit in with Genesis and creation?" Jon stared at Bert.

"Well first don't drag God and Genesis into this yet, you've got to settle the issue of whether Neo-Darwinism is able to explain what we see in the fossil record or not. If not then it's time to investigate other explanations. So is Neo-Darwinism able to explain the top down structure of the fossil record we see in the Cambrian and Ediacaran periods?"

Ellie was looking worried; Jon leaned forward and replied, "Ediacaran period?"

"Yes, there was a mini burst of new phyla and life in less than 40 million years in the Ediacaran period. Prior to that there is nothing but bacterial fossils," Bert replied a bit slowly. "But then in a 5 to 10 million years burst of creative activity 16 more phyla of life popped on the scene during the Cambrian period following the Ediacaran period with no trace of a real evolving precursor.

"You mean evolved don't you," apparently Jon was not joking.

"From what, bacteria? In 5 million years?" Bert snipped back.

"Jon was thinking, and he did not look happy. "Do you know what you are saying?"

He replied.

"Yes, I think I do, some intelligent agency, intelligent design must have been responsible and . . ."

But Jon quickly cut him off, "You are invoking a miracle, a supernatural intervention in the fossil record, science has to throw up its hands and give up, and let the theologians figure it all out with their Biblical texts!"

Ellie looked like she was in anguish, Mary was looking around with a how did I get into this conversation look. Bert answered quickly, "What is so wrong with a miracle? Couldn't God raise a man from the dead?

"It just doesn't fit, Nature is a closed system, and every effect has its cause. You don't need miracles to explain how it all works and fits together. It's just not scientific!" Jon was getting a bit perturbed. Ellie's expression was moving from anguish to horror.

"Jon, calm down, consider your thoughts"

"And what is wrong with my thoughts!" Jon replied.

Bert glanced nervously at Ellie and Mary, "Nothing! Nothing is wrong with your thoughts; just consider the fact that you have them.

"I have thoughts of Ellie all the time!" with that he turned and beamed at Ellie. Ellie was turning a deep shade of red while Mary giggled.

"Well of course, but I am talking about your rational thoughts processes."

"What do you mean?" queried Jon.

Bert started in, "Consider the fact that you have them. Suppose you had a piece of bone pressing in and messing with your brain. Then your thoughts might be irrationally influenced by that injury. We could not trust that the thoughts you uttered were really you and could be trusted, or just the random sayings of a lunatic or madman."

"Well that would explain a lot of Jon's comments," giggled Mary.

Jon rolled his eyes. Bert continued on unfazed, "In the natural world every cause has an effect. Every action has a consequent that logically follows, not always predictably, but follows none the less, unaffected by human choice. But thoughts are not like that. We can choose whether or not we wish to act on our thoughts, it is not a logical necessity that our action will follow a given thought. Very unlike a rock that is pushed over the edge of a hill, it will roll down whether its wants to or not."

Jon stared at Bert, "I wasn't aware that rocks could choose."

"Precisely!" said a smiling Bert.

"Weird, continue."

"It is like our thoughts come from another realm, a realm that can affect the Natural world, but the Natural world can only affect our rational thought world by messing it up. The moment a person's thoughts are thought to have arisen as a result of material causes, like a brain chemical imbalance, or traumatic brain injury, we discount it. Because conscious rational thought does not arise from material causes; it is a volitional action, not a necessity arising from natural law. Thus how can a material process such as Neo-Darwinism have produced conscious rational thinking creatures such as us?"

"Did you dream this up or get it from somewhere?" demanded Jon.

"I have your orders here," the tightly dressed waitress proceeded to place the orders around at the appropriate stations in the booth. She was all business, as everyone paid up, entranced by their fries or Coke. Jon had demolished half of his pizza slice before the waitress bounced away.

"No, I got this from C. S. Lewis, in his book, *Miracles*[22], wish I had dreamed this up," Bert continued. "In fact Lewis phrased it like this . . ."

Mary leaned over and whispers to Ellie, "Here comes another quote!"

Bert continued on, *"Nature can only raid Reason to kill; but Reason can invade Nature to take prisoners and even to colonise. Every object you see before you at this moment– the walls, ceiling and furniture, the book, your own washed hands and cut finger-nails, bears witness to the colonisation of Nature by Reason: for none of this matter would have been in these states if Nature had had her way."*[23]

"This is crazy! Are you expecting me to believe that everything in this restaurant has its origins in the supernatural? This is your proof of the supernatural?! Jon was turning slightly red as he leaned forward to answer Bert.

"I did not say proof, just an argument for a realm of thought that stands outside of Nature, a realm that may have its origin in the original source of thought, God or the above realm (supernatural) that God occupies."

"This is nuts!"

"Show me where I am wrong." replied Bert softly.

"I don't have the time!" With an angry snatch of his remaining pizza, Jon stormed out of the Chicken Palace.

Mary was stunned, "What is wrong with Jon!?" Ellie was trying not to cry.

"My God, I am sorry Ellie, I had no idea he would take it like that," stammered Bert.

"It's not your fault Bert, I think there is a lot going on inside of him," sniffed Ellie, "He has been going with me to church for the last month. I thought I was getting somewhere." Sadly the group finished their snacks, and left to go separate ways.

22. Lewis, Miracles.

23. Lewis, Miracles, 26.

10

Macroevolution

I WAS SEVERAL MONTHS into my second sabbatical stay at Eotvos Lorand University, Budapest, Hungary. I had made friends in the department hosting my stay, and friends with another group of Hungarian academics outside the university. They were the ones who told me of the upcoming induction of Professor John Polyani as an honorary member of the Hungarian Academy of Science. The Academy of Science building was a huge imposing stone structure, near the Danube River; that was just a few tram rides away from my university and Bela Bartok Avenue where my family and I lived. The Academy of Science building was one of those very impressive turn of the century buildings that told of a far more glorious past, that Budapest used to enjoy. A past where Budapest was often called the "Paris of the East." Michael Polanyi's writings are preoccupied with the dramatic fall from this past caused by the triplicate disasters of WWI, WWII, and the Soviet and Eastern Bloc occupation. He watched it unfold in front of him in Budapest, and later Germany, before his family fled to England for safety. Polanyi had a minor problem; he was born of Jewish parents. That did not bode well for him, irrespective that he was now Christian, in the Europe of 1935—1945. With his wife Magda he had two boys, John and Charles. They were raised in England after fleeing Germany and Hungary in their boy's early childhood. Michael Polanyi's son John was raised and educated in England becoming a physical chemist like his Dad. John progressed on to obtain a professorship at the University of Toronto, and later won a Nobel Prize in Chemical Kinetics

239

in 1986. It was in honor of this that the Hungarian Academy of Science made him an honorary member in 2006. How could I pass up a chance to see this famous son of the late Michael Polanyi? So I headed out to the public ceremony at the Academy building that early spring afternoon.

I entered the huge multistoried stone edifice and realized I had not a clue as to where the induction ceremony was to take place. Seeing no one around, nor any stream of people heading anywhere, I carefully choose a grand sweeping stone stairway up as my most likely path. The building was impressively grand and had been carefully renovated. As I climbed about one story up I hear below on the stairs,

"Does anybody around here speak English? How am I going to get to the lecture?" The voice belonged to a middle-aged red headed woman with a very distinct British accent. She had just turned the corner of the stairway enough to see me.

"I speak English ma'am."

"Oh thank God, somebody who speaks English," she replied back to me. "I am trying to make it to my husband's lecture and I don't know what room it is in."

"Which lecture are you going to," I replied.

"My husband is John Polanyi and he is being inducted into the Hungarian Academy here today."

"Oh, I am going to the same lecture!"

"Do you know where it is, can I follow you?" she asked expectantly.

"Sure. I am not sure but I believe the lecture is up here," it was then that I realized that I had the daughter-in-law of Michael Polanyi walking beside me. Here was a chance to learn something about the man from someone who might have known him well. Thus being the slightly idiotic American that I sometimes am, I stated, "I am a great fan of Michael Polanyi's writings . . ." She gave me a slightly irritated look of a woman who was focusing on "her" husband's accomplishments at the moment, and not those of her dead Father-in-law.

"Everybody reads him!" she spat out.

Realizing my faux pas, I changed the topic, and we arrived quickly at the correct lecture hall. But in spite of my social miscue, it was fascinating to realize that even the wife of a Nobel Laureate can be intimidated by their dead Father-in-law; and just who is *everybody*? How far does Michael Polanyi's influence go?

At this point, the reader may notice an overall direction in this tour of science and religion. First we started with an introduction to philosophy and worldviews, and then moved to the unique, but compelling description of science as expressed by Michael Polanyi. From there we started with cosmology, then the solar system, the early earth, chemical evolution, biochemistry, molecular evolution, very early life, paleontology, and now finally to macroevolution. It is like we started way below the Darwinian tree of life and then started working our way up until we reached the very start of the tree of life, and then started working our way up it. This involves starting from a physics, chemistry and geology background to analyze the claims of Darwin, long before we invoke biology. When the grand metaphysical story of Neo-Darwinism is examined in this way first, the major flaws and problems become immediately apparent. It is from this context that the biological claims can be examined with a different eye. Most treatments of the subject tend to take a biology first approach, and then delve into the other areas. Starting at the top of the tree of life and then working their way down to the bottom. Neo-Darwinism works best at the top of the tree of life, and as we have noted has massive seldom discussed problems at the bottom of the tree of life. Enough even to question the whole concept of a tree of life. Thus it is like two opposite trends are at work. Negative evidence and tremendous unsolved problems that fly in the face of what the hard physical sciences are willing to tolerate, predominate at the bottom of the tree of life. At the top of the tree of life is where you find the best case for and the most evidence for Neo-Darwinism. Thus somewhere in the middle is where the two conflicting strands meet, and you find what Professor Michael Behe calls the "edge of evolution."

The diagram below, the classic tree of life, graphically illustrates the point. Intelligent design arguments strongly invalidate the bottom of the tree of life. Question marks are drawn near the top of the tree of life and are drawn with increased frequency as one goes down the tree in time, indicating the much lower degree of confidence and certainty associated with the data and its interpretation. I was first introduced to this concept in an influential book, *How to Think About Evolution and Other Bible-Science Controversies*[1], by L. Duane Thurman, that I read in my college days. The modern biologist will quickly point out that the classic tree of life is not valid anymore. Modern depictions use bushes, webs or other schemes. But that is a telling point itself. The issues raised by Cambrian explosion and molecular homology have wreaked havoc with the traditional tree of life, and one must retreat back to traditional homology to get any sort of tree of life you can draw or easily discuss. This will be discussed later.

---

1 Thurman, How to Think About Evolution and Other Bible–Science Controversies.

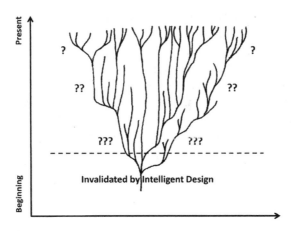

Somewhere in the question marks is the "edge of evolution." Different people will move that edge of evolution up or down the tree of life depending on their training and philosophical worldview. The diagram below comes from a talk the author gave at an annual meeting of the American Scientific Affiliation[2]. It attempts to explain from my own personal perspective and experience why different fields of science approach the evolutionary tree-of-life explanation with such different degrees of confidence.

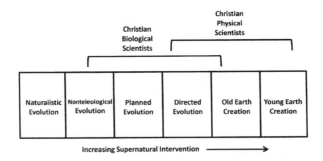

In this diagram you will notice that scientists tend to be spread along a continuum ranging from young earth creationism (YEC); to old earth creationism (OEC); to directed evolution (DE)—God or a higher agency intervened in some fashion to guide and direct evolution via non-random or non-stochastic processes; to planned evolution (PE)—God set up the laws and processes at the front end of any evolutionary process so that we cannot detect His action; to non-teleological evolution (NTE)—God or some higher agency exists but he had little to nothing to do with the natural laws

2. Collier, "The Necessity of a Polanyian Perspective."

of the universe and hence the course of evolutionary action; to naturalistic evolution (NE)—God does not exist and the laws of nature are sufficient to explain the origin of the world and life, i.e. atheism or agnosticism. The NE, NTE, PE, DE, OEC, YEC part of the diagram comes from similar figures and discussions in Gerald Rau's *Mapping the Origins Debate*[3].

The YEC's tend to buy very little of evolution, the OEC's just part of it. DE's buy portions of evolutionary theory but highly doubt it creative powers. PE's and NTE's tend to accept most if not all of it; but NE's, who are usually agnostics and atheists, tend to accept the whole *molecules to man by random chance* package totally. It usually functions as their creation story and is crucial to the coherency of their worldview. There are many exceptions of course, but these trends are fairly faithful. Inspection of the above diagram reveals a significant shift of the "hard" physical sciences scientists (physicists, chemists, geologists, mathematicians) to the right (or less confident in evolution), from the biological sciences. My proposed explanation comes from the two different paradigms that the two groups of scientists indwell, or use. First let's define a secondary meaning of a rarely used word; physiognomy—the general form or appearance of something. Now look at this quote from Michael Polanyi.

> We have seen that living beings are characterized by their physiognomies, including the space-time physiognomy of their functions. A physiognomy obeys no mathematical formula; it can be recognized only tacitly by dwelling in its numberless particulars, many of them subliminal. We can know a frog only by dwelling in its particulars in this way, and we can know the topography of a frog only if we are able first to know a frog.[4]

A scientist indwells and uses his scientific training and theories to discover new novel things much like a blind man uses a cane to find his way around. The sounds of the cane tapping and the feel of the push back on the blind man's hand give him myriad clues as to the ground in front of him. He indwells the cane to "*see*" his way along to new territories. The scientist indwells his taught and learned paradigms/theories to feel his or her way along to new discoveries. To the physical scientist who indwells thermodynamics, classical mechanics, electrodynamics, quantum mechanics, etc., to understand his scientific world, the production of life from inanimate matter must be viewed from such a perspective, and it colors the view of the whole. The clash between chemical thermodynamics and chemical evolution is critical. The biologist first learns the Linnaeus classifications and then

3. Rau, Mapping the Origins Debate.
4. Polanyi, Meaning, 143.

the Darwinian perspective; indwells their theories to "*see*" their way to new discoveries. Such a clash has little meaning to them unless viewed through the lens of evolutionary theory.

Questions like, "Do you believe in evolution?" are a bit simplistic and laden with the logical problem of the *excluded middle*. The problem of the excluded middle comes from framing a problem in terms of two different polar extremes and ignoring the fact that intermediate positions are possible. You are forced by the question to answer in terms of extremes or as the question asker has framed the problem. The best strategy in such cases is to redefine the question on your own terms before answering, or ask why the questioner has phrased the question in that manner.

I am a "hard" physical scientist, a physical chemist; a molecular spectroscopist with extensive training and experience in thermodynamics, quantum mechanics, statistical mechanics, molecular spectroscopy, math and probability theory, chemistry, biochemistry, and how they relate to the biological sciences. These fields overlap with Neo-Darwinism in significant ways, particularly at the bottom of the tree-of-life, and before. That training has been a significant impetus to the writing of this text, and it would be improper to not state that I have been heavily influenced by the hard sciences paradigm and my Christian worldview. Thus how should I discuss the more biological aspects of evolution which are moving outside my field of experience?

One perspective is to say that the field is outside my field of expertise and defer to biological experts. Thus only experts should be allowed to talk about their respective fields, and those of us who are not experts had better keep our mouth shut. But instinctively most of us realize this is laden with problems. How do you define an expert in a field? What do you do when experts within a field disagree? What if an expert in a field speaks authoritatively over an area in his field that overlaps with things outside his field, that you know from your experience and training are just flat wrong? Do you just stand by and let the error persist just because *part* of what he is talking about is outside your domain of expertise? Or do we as scientists want our respective disciplines to be so clouded in high level expertise and jargon that the average taxpayer who is footing the bill for our research has no idea of what is at stake? What are our goals, our progress, and why they should continue to fund us?

Better yet; they should become so acquainted with what we do that they become excited over our new discoveries. Even if they don't have the training to do it as we do. That is the positive side of making our professions accessible to the layman. But with it comes the price that they can ask more penetrating questions, so of which are uncomfortable. Biology, (and macroevolution in particular), is in the enviable position of being much more accessible to outside scrutiny. Something that I wish there was more of in my own field of Molecular Spectroscopy. Someday I will go to a community-wide symposium on "Is Quantum Mechanics the Final Solution for Spectroscopy?," or "Why Should my Child be Taught Electromagnetic Theory at such an Impressionable Age"? Until then I remain envious.

The route chosen here is to assume that many readers, have been, or will be, exposed to significant teaching in favor of an evolutionary view, usually from a materialistic or random chance mechanism that drove it from molecules to man. Secular western culture often adds that God may or may not have been involved, but if God was, it had to be at the very beginning, and it is impossible to scientifically detect or theorize about it. So I will present a short list of arguments against macroevolution, which are rarely discussed in the public educational square, and let the reader decide for themselves how much of macroevolution or evolution in general, they wish to accept or deny. My personal preferences will probably show though in spite of attempts to be objective, but that is the price one must pay, if you want to say anything significant.

First let's start with a list of unsolved problems of macroevolution.

1. *The origin of life—Where did the genetic information come from? How did it evolve or originate? What mechanism? How did nonliving material matter become living?*

2. *The origin of biological information and its increase in the tree of life*

3. *The development of irreducibly complex molecular machines*

4. *The biological big bang of the Cambrian period*

5. *The origin of multicellular life—How does the cell know which way to develop as it splits and reproduces?*

6. *The origin of sexuality*

7. *The scarcity of transition forms in the fossil record*

8. *The development of complex organ systems*

Let's start with this list. First:

The origin of life—Where did the genetic information come from? How did it evolve or originate? What mechanism? How did non-living material matter become living?

We have spent extensive time and explanation dealing with this topic in *Chapter 5—Cells, Machines and Biochemistry* and *Chapter 6—Primordial Soup.*

The origin of biological information and its increase in the tree of life

This has been dealt with in *Chapters 6—Primordial Soup, Chapter 8—Change and Evolution* and *Chapter 9—Cambrian Explosion and the Tree of Life.*

The development of irreducibly complex molecular machines

We discussed this in *Chapter 8—Change and Evolution.*

The biological big bang of the Cambrian period

This was covered in *Chapter 9—Cambrian Explosion and the Tree of Life.*

The origin of multicellular life—How does the cell know which way to develop as it splits and reproduces?

This is an interesting problem. All of us started out life as a single fertilized egg/cell in our Mother's womb. As this cell divided and reproduced, we became a multicellular creature. Some of these cells became our brain, some our spines, some our heart, and some moved all the way down, and became our toes and toenails. How did they know to do this and what source of information guided their placement and role? In the educational public square, proteinomics, or other chemicals are said to guide embryonic development, but the details are extremely murky, because when it is all said and done, we really don't know. The quantity of information, blueprint construction information wrapped up in this process, may likely greatly exceed the protein sequencing and construction information already stored in our DNA. Is it possible to store this type of information in our DNA? Is it even there? How did it come to be? How in the world, (or better yet, why in the world), did a series of random micro-mutations get selected to produce this? How could natural selection which has no mind of its own, just blind stochastic processes, construct a blood pumping heart, or fin flapping limb out a mess of reproducing cells of the same type that had nothing of the sort to start with. No one really has a serious clue.

The origin of sexuality

Sex is an interesting problem. How did it originate? We know some simple creatures and plants can reproduce asexually, they just bud off; and others use sexual means where the male passes genetic information on to the female before reproduction occurs. Some creatures can switch between both methods as the need arises. But the information gap between the two methods of reproduction is profound and deep. Thus how does a very slight incremental change in organism characteristics generate a completely new and novel means of reproduction? Particularly when it is very hard to envision a series of reproductive systems that grade from asexual to completely sexual where every single one of them must work or the creature fails to reproduce. How do you get half a sex? In particular how do you get half a sex, (or 3 quarters, or 0.00246 sexual) that actually works so that natural selection can select for it? And choose it over a fully functional asexual means of reproduction; since some means of reproduction must be functional or the whole evolutionary improvement grinds to a halt. Building an evolutionary bridge from asexual to sexual where every step or increment of the bridge must work and offer a functional advantage for natural selection to work on, boggles the imagining mind. And if it actually occurred where are the transitional systems that still ought to linger; particularly since they had to be evolutionarily successful to get the creature from one side of the asexual-sexual bridge to the other.

The scarcity of transition forms in the fossil record

This has been covered some in *Chapter 9—Cambrian Explosion and the Tree of Life*, buts merits further discussion here. Earlier and in chapter 9, I mentioned that the fossil record is marked by stasis and the sudden appearance and disappearance of new animal and plant species particularly in the Cambrian period. But this is also true of almost all of the fossil record. Here is a quote from Phillip Johnson[5] in *Darwin on Trial* where he is also quoting

---

5. Johnson, Darwin on Trial, 50.

the founders of punctuated equilibrium, Stephen Jay Gould, Niles Eldredge, and Steven Stanley.

> The history of most fossil species includes two features particularly inconsistent with gradualism:

> 1. Stasis. Most species exhibit no directional change during their tenure on earth. They appear in the fossil record looking pretty much the same as when they disappear: morphological change is usually limited and directionless.

> 2. Sudden appearance. In any local area, a species does not arise gradually by the steady transformation of its ancestors; it appears all at once and "fully formed."

The question to ask is how well has this aspect of the fossil record been publicized? Johnson quotes David Raup of the University of Chicago and the Field Museum. Raup is one of the world's most respected paleontologists.

> A large number of well-trained scientists outside of evolutionary biology and paleontology have unfortunately gotten the idea that the fossil record is far more Darwinian than it is. This probably comes from the oversimplification inevitable in secondary sources: low level textbooks, semi-popular articles, and so on. Also, there is probably some wishful thinking involved. In the years after Darwin, his advocates hoped to find predictable progressions. In general, these have not been found—yet the optimism has died hard, and some pure fantasy has crept into textbooks . . . One of the ironies of the evolution-creation debate is that the creationists have accepted the mistaken notion that the fossils record shows a detailed and orderly progression and have gone to great lengths to accommodate this 'fact' in their Flood geology.[6]

---

6. Johnson, Darwin on Trial, see research notes Darwin on Trial, p. 186; Raup's quote was pulled from Science, vol. 213, p.289. Interestingly, Raup reiterates and expands on this idea even more:

For this reason alone, stage of evolution could not be used to build a geologic time scale. The whole problem is made more difficult by the fact that a surprising number of geologists with specialties other than paleontology share the same misconceptions. Wysong takes obvious pleasure in quoting W. M. Elsasser in the Encyclopaedia Britannica (1973) as saying, "the geological method presumes the existence in these periods of living beings of gradually increasing complexity" (1976, pp. 352–53). Professor Elsasser is an excellent geophysicist, but his expertise in fields distant from geophysics cannot be expected to be optimal. The creationists (and probably Professor Elsasser) come by their misunderstanding honestly, at least in part. Many teachers and textbook

Granted I could be accused of quote mining, because David Raup is certainly not in favor of Johnson's evolutionary questioning, and it is doubtful he would support much this chapter and text. But if we want to illustrate the reality that the fossil record really is very discontinuous, out of sync with Darwin's gradualism and many popular depictions of the fossil record; the quote is very compelling. We finish with one more quote from Raup, as mentioned by Johnson in *Darwin on Trial*, but it comes from the book, *Scientists Confront Creationism*.

> Darwin predicted that the fossil record should show a reasonably smooth continuum for ancestor-descendant pairs with a satisfactory number of intermediates between major groups. Darwin even went so far as to say that if this were not found in the fossil record, his general theory of evolution would be in serious jeopardy. Such smooth transitions were not found in Darwin's time, and he explained this in part on the basis of an incomplete geologic record and in part on the lack of study of that record. We are now more than a hundred years after Darwin and the situation is little changed. Since Darwin a tremendous expansion of paleontological knowledge has taken place, and we know much more about the fossil record than was known in his time, but the basic situation is not much different. *We actually may have fewer examples of smooth transitions than we had in Darwin's time, because some of the old examples have turned out to be invalid when studied in more detail.* To be sure, some new intermediate or transitional forms have been found, particularly among land vertebrates. But if Darwin were writing today he would still have to cite a disturbing lack of missing links or transitional forms between the major groups of organisms[7].

(Emphasis added by Phillip Johnson)

This quote comes from the 1983 first edition of the book *Scientists Confront Creationists*. Raup goes on to explain that the lack evidence for

---

writers, especially in the late nineteenth and early twentieth centuries has been so carried away by the elegance of the Darwinian model that they have ascribed powers to it that do not exist. It would be a fine thing if we could use some abstract estimate of stage of evolution to date rocks—but we cannot!

An interesting irony in this whole business is that the creationists accept as fact the mistaken notion that the geologic record shows a progression from simple to complex organisms. (Laurie Godfrey, Scientist Confront Creationists, W. W. Norton and Company, New York, New York, 1983, p. 154-155.)

7. Godfrey, Scientists Confront Creationists, 156; and Phillip Johnson, Darwin on Trial, 187.

transitional species can be explained because of 1) the nature of the classification system creatures have to be put in one group or another, and so their absence is an artifact of the classification system; 2) the fossil record is still incomplete; 3) evolution may occur rapidly by punctuated equilibrium. Raup's conclusion: "With these considerations in mind one must argue that the fossil record is compatible with the predictions of evolutionary theory." Johnson mentions in his research notes of *Darwin on Trial*, pages 156–158; "I think that the phrasing of that conclusion hints at a certain lack of conviction."

What is fascinating is to examine the 2007 second edition of *Scientists Confront Creationists: Intelligent Design and Beyond*, by Andrew J. Petto and Laurie R. Godfrey, eds.[8] The chapter on transition specie in the fossil record is still included, but it has been redone by other authors. Instead of David Raup; Kevin Padian and David Angielczyk write the chapter and explain away the lack of transitional specie with 1) the fossil record is incomplete; 2) The rise of phylogenetic systematics has changed the classification from differences and similarities to classifying according to genealogical order; 3) shifting the focus from transitional *forms* to transitional *features*; 4) phylogenetic analysis also helps reveal the sequence of changes that forged new adaptations from pre-existing structures. *Gone is any mention of punctuated equilibrium in the 2007 conclusions.* Punctuated equilibrium evolution was one of the main points used for matching Darwinian evolution with the fossil record in the first 1983 edition of *Scientists Confront Creationists*. This is very telling; and suggestive of the negative shifts in the paleontological community over the explanatory powers of punctuated equilibrium on the fossil record. If you examine the four points used to explain the lack of transition specie in the 2007 edition you find– Point 1) the fossil record is incomplete: no serious student expects the fossil record to be complete. The real question is whether it is a representative sample of the complete fossil record. As mentioned earlier in chapter 9, there is serious evidence to suggest it is representative. Points 2) and 3) of the 2007 edition seem to look at the existing fossil record in another way; by looking for transitional *features* instead of the actual expected transition *forms* (fossils?). The fossil record hasn't changed; we just are *looking* at it differently. Point 4) is quite fascinating. Phylogenetic analysis requires genetics, or DNA, or proteins and stuff from which to analyze these genetics. This requires looking at present day modern living species and studying the differences in their proteins and DNA to build up evolutionary trees and inferred transitional forms from the calculated (imaginary or real nodes on the tree?) phylogenetic distances.

8. Petto, Scientists Confront Creationists, 198–199.

In my opinion this looks like a significant retreat from actual examination of the fossil data to more obscure computer oriented treatments of the pre-existing data.

Before we leave transition species it worth looking at one fairly famous sequence of fossils reported to show the transition from reptiles to mammals. These diagrams come from Meyer, Minnich, Moneymaker, Nelson, and Seelke in their book *Explore Evolution*[9]. The left sequence has its source T.S. Kemp, *The Origin and Evolution of Mammals*, 2005. The right sequence fossils are shown at their correct relative size as calculated by the authors of *Explore Evolution*. The *Explore Evolution* authors also point out that the fossil skulls did not come from the same fossil bed, or even the same nearby geographical location. Some paleontologists might protest that demanding that all sequence fossils come from the same fossilized beds is asking too much. But is it too demanding to ask that these facts be included with the diagram in addition to noting the renormalization of the skull sizes?

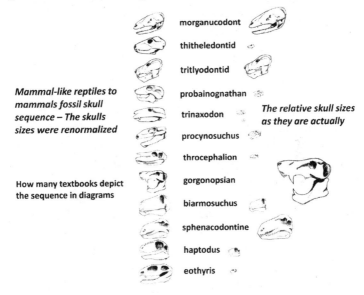

Now let's examine the next macroevolution problem.

9. Meyer, Explore Evolution, 21, 29. The left side diagram is loosely based on the diagram on page 21 and the right side likewise from page 29. The left side diagram is sourced from T.S. Kemp, The Origin and Evolution of Mammals, Oxford: Oxford University Press, 2005, 89.

The development of complex organ systems

Shown below are diagrams of a reptile three chambered heart and a mammal four chambered heart similar to diagrams from *Explore Evolution*[10].

Three chambered reptile heart

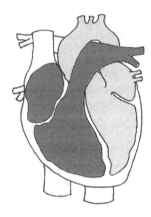

Four chambered mammal heart

  The lower ventricles are not completely separated and oxygenated and un-oxygenated blood mix a bit during the pumping process. In addition the pulmonary arch and the aortic arch of the reptile heart is plumbed significantly different from the mammal heart pumping system which completely separates the oxygenated blood on the left side from the un-oxygenated blood pumped on the right side of the heart. To effect an evolutionary transition from a reptile to a mammal means much more than transforming their lower jaw bones; it means re-plumbing the heart and circulatory system completely from one setup to the other—all while keeping the heart and circulatory system functioning effectively during every single step of the transition. We need a working set of transition systems that make sense, for which there is "some evidence" that it happened. This is rarely discussed if at all.

  These are diagrams for a reptile breathing system and a bird or avian breathing system also taken from *Explore Evolution*[11].

  The left diagram shows the classic in and out system of reptiles; the lung expands, air comes in, goes through the bronchial tubes and tiny air sacs, the alveoli, where the blood flow is oxygenated. Lastly the exchanged

10. Meyer, Explore Evolution, 131.

11. Meyer, Explore Evolution, 133–7.

carbon dioxide in the alveoli is pushed out of the lungs when the lungs are compressed and the stale air is pushed out. The right side diagram, which is greatly simplified, shows that bird or avian lungs are completely different. The air is pumped into the lungs and alveoli, (oxygenating the blood), via a dual expansion and compression of the ribcage and then expelled out to a *completely different location* via another series of compression and expansion of the ribcage. The remarkable details are given in *Explore Evolution*[12]. How do you transition from one lung system to another without killing the creature? The details and possible scenarios are very lacking.

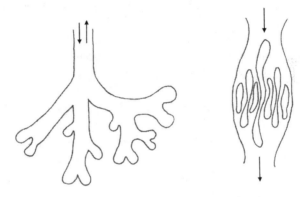

Reptile in and out lung   Bird flow through lung

    Let's try to envision a good mechanical analog to both the heart transformation and the lung transformation problem. Take an expensive Harley-Davidson motorcycle and start it up; now let's change it to a four door Mercedes Benz with a V-8 engine instead of the Harley V-2. You can setup a nearby factory if you wish to build and bring in the needed parts and intelligent engineers for the transformation. But there is one stipulation that you must always follow. The motorcycle to motorcycle-car to car must be kept running the entire time of the transformation. If it stops for even one second, you are disqualified and it is over. By the way; you also need to do this with blind random stochastic processes. Ready to go? How much money do you want to bet on the success of this transformation? Granted these machines do not reproduce like living things; but the engineering problems are of the same magnitude if not worst. How does a blind stochastic process invent such a transformation?

    The last topic is a problem not listed earlier in this chapter, but ties into macroevolution and the biochemical chapters that preceded it; using

12. Meyer, Explore Evolution, 133–7.

molecular phylogenetics to reconstruct the universal tree of life. In 1962 biochemist Emile Zuckerkandl and physical chemist Linus Pauling suggested that DNA sequences could be used to find or justify evolutionary trees. The basic idea is that that by comparing the DNA sequences of different animals you could see how "similar" or "dissimilar" they are and then calculate a quantitative measure of how "related" the two animals are. Then by using the measures of relatedness a phylogenetic or evolutionary tree of life could be constructed. If it matches up with tree of life produced by classical biologists using homology and morphological similarities, then you had a convincing case for the Darwinian tree of life. It was not long before scientists realized that they were not limited to DNA sequencing. They could use common types of RNA, and proteins that are common to all or many organisms, like cytochrome c, cytochrome b; even microRNA genes were fodder for the phylogenetic mill.

Initially RNA and cytochrome c were used with some measure of success. True that some tree divisions had to be rearranged from the classical homology based trees, but overall there seem to be some sort of matchup. But as scientists starting expanding into other proteins, genes, microRNA's, a disturbing trend was noticed. Different proteins, different genes, would give different trees of life, sometimes radically different trees. Dartmouth biologist Kevin Peterson wrote,

> I've looked at thousands of microRNA genes and I can't find a single example that would support the traditional tree . . . a radically different diagram for the mammals: one that aligns humans more closely with elephants than with rodents . . . The microRNAs are totally unambiguous . . . they give a totally different tree from what everyone else wants.[13]

From an article in Trends in Ecology and Evolution,

> The mitochondrial cytochrome b gene implied . . . an absurd phylogeny of mammals, regardless of the method of tree construction. Cats and whales fell within primates, grouping with simians (monkeys and apes) and strepsirhines (lemurs, bushbabies and lorises) to the exclusion of tarsiers. Cytochrome b is probably the most commonly sequenced gene in the vertebrates, making this surprising result even more disconcerting.[14]

Finishing with a final quote from a review article in *Nature*,

---

13. Dolgin, "Rewriting Evolution," 460–2.

14. Lee, "Molecular Phylogenies Become Functional," 177.

"disparities between molecular and morphological trees" lead to 'evolution wars" because "evolutionary trees constructed by studying biological molecules often don't resemble those drawn up from morphology."[15]

All of these quotes were drawn from Casey Luskin's chapter, "The Top Ten Scientific Problems with Biological and Chemical Evolution," in the volume *More than Myth*[16]. There are several more revealing quotes along with Luskin's discussion in the reference given.

What does this mean? Basically the molecular evidence for a tree of life varies depending on the key molecule used, and it doesn't really agree with the homology derived tree of life. It could be that Zuckerkandl and Pauling just hypothesized wrong or else they and others have missed some unifying factor that ties the molecular data into a unified tree of life. It could be that they were viewing the molecular data with the wrong paradigm. Could it be that the molecules are not related because of evolution and common ancestry; but related because of "similar" or "dissimilar" design constraints? Given our current scientific philosophical environment, will alternate hypotheses ever be investigated? For example, shown next is a diagram created by Ken Weed, a college Dean and fellow professor. He frequently presented this diagram[17] to students and queried them as follows. On the left side of the table is a listing of animals/creatures; starting with human at the top, then chimpanzee, etc., down to frog, then down to deer tick to mosquito, and so on, ending at Baker's yeast. Quite a spread of creatures; and they all use cytochrome c protein as a necessary protein for their metabolism. Now read across from human left to right and you will see a string of letters. Each letter represents one of the 20 possible amino acids that are present in each amino acid position in the cytochrome c protein chain. Cytochrome c for humans starts with MG————DVE and then continues. The dashes—represent an amino acid that is missing. Thus when we go to the bottom of the table like for deer tick we see MGDIPKGDPE. The dot . means simply that the amino acid is the same at the same position as in human cytochrome c. Thus by scanning down the chart we can easily see all the differences between the cytochrome c for all the creatures. If it has a dot . then the amino acid is the same as in the human cytochrome c protein. If it has a letter then that particular amino acid at that position is different and is given by the

15. Gura, "Bones, Molecules or Both," 230–3.

16. Luskin, "The Top Ten Scientific Problems," 8.

17. Used by kind permission of Dr. Ken Weed, Dean of the College of Science and Engineering, Oral Roberts University, Tulsa, Oklahoma, 2018.

respective letter. If it has a dash—then it just does not have that amino acid at all instead of a substituted amino acid.

Looking at the chart we can quickly see that humans and chimpanzees have identical cytochrome c proteins. We can compute the differences between the different species and then from a measure of their differences we could try to build an evolutionary tree. Or we can stare at the cytochrome c table and ask, "Do I see any significant pattern differences? And why are they there?" Usually when the data is presented this way, students quickly notice all of the insects have the pattern VPAG near the start of the protein sequence, except for the honeybee which has IPAG. Is this because of evolutionary distance and common ancestors or because the cytochrome c in flying insects requires an additional structural function in the protein that is necessary for their unique role in the ecosphere? Is it *descent with modification* that caused this difference or needed *design constraint*? It all depends on your perspective. How will we know if one is better than the other, if all the research money is spent investigating only one perspective?

Cytochrome C Amino Acid Sequences

Name	#	Sequence
Human	1	MG-----DVE KGKKIFIMKC SQCHTVEKGG KHKTGPNLHG LFGRKTGQAP GYSYTAANKN KGIIWGEDTL MEYLENPKKY IKKKEERADL IAYLKKATNE
Chimpanzee	2	..-----... ....
Chicken	3	..-----.I. ...VQ.. ....E .F...D.... ...T...... ....S..V... .....D..SK
Cattle	4	..-----... ...VQ.. A......... .... .F...D.... ...T...E.. ....G..E... ....
Horse	5	..-----... ...VQ.. A......... .... .FT..D.... ...T.K.E.. ....T..E... ....
Mouse	6	..-----... ...VQ.. A......... .... .F...D.... ...T...... ....G...... ....
Monkey	7	..-----... ...V.VQ. A......N.. ...V....W. ....E .F...D..S ...V...... ....G...... ....S..S.
Tarsier	8	..-----... ...VQ.. A......... ......?.... ....A .F...D.... ...T...... ....G...... ....
Pig	9	..-----... ...VQ.. A......... .... .F...D.... ...T...E.. ....G..E... ....
Frog	10	..-----.A. ...VQ.. ....E .F...D.... ...V.D.G.. L........ ...G..Q... .....QS.SS
Deer Tick	11	..DIPKG.P. ...V.VQR. A......A.. ..........I...S. .F...D...S ...T.NK... FI...D.... L..VQD.... MS..EQ..K-
Honey Bee	12	..-IPAG.P. ...VQ.. A...I.S.. ...V....Y. VY........ ...D...G ...T.NKE.. F........ L..PQ..... ...IEQ.SK-
Fruit Fly α	13	..-VPAG... ...L.VQR. A......A.. ...V...... .I........A .FA..D...A ...T.N.... F........ L..PN..G.. .....S..K-
Fruit Fly β	14	..-VPAG... ...L.VQR. A......A.. ...V...... .I........A .FA..D...A ...T.N.... F........ L..PN..G.. .....S..K-
Flesh Fly	15	..-VPAG... ....VQR. A......A.. ...V...... .... .FA..D...A ...T.N.... F........ L..PN..G.. .....S..K-
Mosquito	16	..-VPAG... ....VQR. A......A.. ...V...... .... .F...D...A ...T.N.... F........ L..PQ..G.. .....S..K-
Worm	17	.ADIPAG.AA ...V.VQR. A......A.. ....S..... .... .F...D.... ...T..K.. WV...... L...N..... .....EES.K-
Fission Yeast	18	.P-YAPG.EK ..ASL.KTR. A......... AN.V...... V.........E .F...E..RD ..T.D.E.. FA....... F..PAD.NNV .T......S-
Baker's Yeast	19	.TEFKAGSAK ..ATL.KTR. L......... P..V...... I...HS..E .....D..IK .NVL.D.NNN S...T..... L..EKD.N.. .T.....CE-

After looking at some of these objections to macroevolution, it would be tempting to discard it all. But that is only one side; it would be more honest to list some of the evidence that does exist to support evolution to try to gain a deeper perspective. Michael Denton is an Australian molecular biologist and medical doctor who wrote the book *Evolution: A Theory in Crisis*. This book caused a lot of professionals to re-examine the claims of modern Neo-Darwinism. Chapter 4 is entitled "A Partial Truth[18]" and there Denton lays down the positive case for evolution as he sees it.

18. Denton, Evolution: A Theory in Crisis, see Chapter 4, 79.

1. Oxford biologist Benard Kettlewell tested the survival rate of a light and dark phase of a woodland moth in dark polluted forests versus lighter unpolluted forests. His comments:

> We decided to test the rate of survival of the two forms in the contrasting types of woodland. We did this by releasing known numbers of moths of both forms. Each moth was marked on its underside with a spot of quick-drying cellulose paint; a different color was used for each day. Thus when we subsequently trapped large numbers of moths we could identify those we had released and establish the length of time they had been exposed to predators in nature.

> In an unpolluted forest we released 984 moths: 488 dark and 496 light. We recaptured 34 dark and 62 light: indicating that in these woods the light form had a clear advantage over the dark. We then repeated the experiment in the polluted Birmingham woods, releasing 630 moths: 493 dark and 137 light. The result of the first experiment was completely reversed; we recaptured proportionately twice as many of the dark form as of the light.[19]

2. Phenomena of circular overlaps. Where a chain of intergrading subspecies forms a loop or overlapping circle, but at the ends of this loop or circle you get two species that do not interbreed with each other. A classic example is the Herring Gull (Larus argentatus) and the Lesser Black Backed Gull (Larus fuscus). In Europe the two species behave and look quite different, and do not interbreed. As you go east across Russia and Siberia the Herring Gull doesn't occur and the Lesser Black Backed Gull starts looking less like itself and more like a Herring Gull. By the time you reach the Bering Straits and Alaska it starts to resemble the Herring Gull. Likewise if you travel west from Europe the Lesser Black Back Gull does not occur but the Herring Gull does and it slowly starts to resemble the Lesser Black Back Gull as you move west. In Eastern Siberia there is a form of the Herring, (or Lesser Black Backed Gull), that is almost exactly intermediate between the two. All of these subspecies interbreed with their adjacent neighboring species or sub-species. But where the terminal ends meet in Europe the two species, Herring Gull and Lesser Black Backed Gull, do not breed with each other at all.[20]

---

19. Kettlewell, "Darwin's Missing Evidence," 148–53.
20. Mayr, Populations, Species and Evolution, 291.

3. There are about 4300 species of insects unique to the Hawaiian Archipelago that appears to have descended from 250 colonizations.[21]

4. The Cape Verde Island Archipelago in the Atlantic Ocean is about 400 miles west of the African mainland. The Galapagos Islands in the Pacific Ocean are about 600 miles west of the South American western Coast. They both have very similar geologies and climates. However the Galapagos has wildlife and plant life that appear to be closely related to South America and the Cape Verdes Island fauna and flora appear to be related to Africa; as if distant ancestors made their way to the island and changed with time.

5. The Wallace-Darwin line exists in the Indonesian Island Archipelago. Northwest of this line fauna and flora closely resemble Southeast Asian species. Southeast of this line the species resemble the remarkable marsupials and flora of Australia. Plate tectonics has revealed that the two areas are moving apart from each other via plate tectonics where the dividing line is right at the Wallace-Darwin line.

What does this all mean? In all of the above mentioned examples, (and there are more examples, ask any evolutionary biologist), the degree of change does not go past the species, or possibly the genus, or family (?) level. Not that far off from Michael Behe's observations and thoughts on where the edge of evolution is. If you are looking at the tree of life from the top it is very easy to extrapolate down and call it a fact of science. Whether or not that "fact" extends to the deepest levels of extrapolation is highly, debatable. If you look at the tree of life from a bottom up view, the case for evolution and the tree of life is pretty terrible. From the physical scientist's perspective it is tempting to call it impossible and extrapolate it all the way up. Somewhere in the middle is the "edge of evolution" and different scientists and different people will have different opinions as to where the edge will be. As a physical scientist who is a physical chemist specializing in molecular spectroscopy, I am pessimistic about evolution; other scientists will differ.

"Hey, there is Ellie and Katrina in the Chicken Palace, let's stop in and see what is going on," Bert and Charles were out and about that warm spring day and had spied the pair of girls in their favorite snacking grill.

Opening the door, Charles called out, "Hey girls, do you need some company?"

Katrina replied, "We have room in the booth, come on in." Charles and Bert slid into the booth across from Ellie and Katrina.

21. Mayr, Populations, Species and Evolution, 291.

"How is it going?"

"Fine."

"Could be better, but doing okay otherwise," was Ellie's reply.

"Have you seen Jon since his last blow-up?" Bert cautiously questioned.

"Not really except at a distant where he just waved at me."

"That's not good," slipped in Charles.

"Maybe yes, maybe no, who knows?" sighed Ellie.

Bert stiffened slightly and said, "Is that him in the distance walking this way?"

Charles craned his head, "By golly I think you are right. Ellie got a panicked look in her eyes, and made leaving motions.

"Ellie don't be silly, stay right here," Bert held out his hand and made staying motions.

Sure enough, it was Jon; he walked down the sidewalk towards the Chicken Palace and then started briefly when he recognized the group sitting inside. With a quick look of serious intent, he came inside Som and Ela's Chicken Palace, strode over to their booth and then slipped inside next to Ellie.

"Hi!"

"hi"

"hello"

"Hi . . ."

"Ellie, I need to apologize to you, (and Bert), for my behavior last time." Jon was speaking quietly to Ellie with a side glance over to Bert.

"Sure," replied a smiling Ellie.

"I was wrong, there really is a supernatural and God reigns in it," Charles's lower jaw was slowly descending to the table, floor and beyond at Jon's remarks. "That the material world is all there is, just doesn't make sense anymore," Jon was looking pensive as he finished his remark.

"What made you change your mind?" asked Bert.

"Well I realized that Ellie and I were imperfect people possibly in need of therapy, and then I read *Keep your Love On!* by Danny Silk.

"Really! Is that what did it?" stuttered Bert.

"Of course not! You twitface! I read C. S. Lewis! *Miracles*, and *Mere Christianity*. What else do you think would do it!" Jon was half angry and half amused. Ellie and Katrina were trying their best not to convulse in laughter. "Lewis is a very persuasive man, very persuasive and quite reasonable."

Bert recovered himself enough to mumble, "Yes, yes a very reasonable man!"

"Do you think you had anything to do with it?"

"Well I was hoping so . . ." whispered Bert.

"Dream on dude. Now Lewis, there is someone who is hard to argue against," Jon was smiling broadly at Bert now. "Actually I have been leaning this way for quite a while, but I just did not want to admit it in public, can't lose face you know."

"Well at least you are there," murmured Bert.

"Have been for a while," replied Jon.

"Jon, why couldn't you have told us, told me?" Ellie was puzzled.

"Ellie, it takes the mystery away. I enjoy keeping people on their toes. It is not like there is this great deciding point where you don't believe, but bang, you see the light and now you do. For me it has been a slow turning away from one perspective to another by degrees and small shifts. If you asked me for the great turning point I could not tell you; does it matter really? This is why it is important to maintain friendships, lines of communication. It helps when you politely bring up valid logical points and aspects to one's worldview. Most people in my opinion are won over by long cumulative arguments, not by dramatic turn-arounds. Sure such turn-arounds happen and are valid, but I personally don't think it is common to most people. Most of us need time to think about things before deciding on worldview changes. Still don't know about this Christianity stuff. Jesus said some troubling things." Jon actually looked a little exhausted; content, but exhausted.

Katrina chimed in, "Well I want something to eat!" And with that, the chatting group set upon the next available waiter.

11

View from the Bottom

THE DATA BELOW IS taken from Table 2.1 in Gerald Rau's *Mapping the Origins Debate*[1].

Six Models of Origins

1. *Naturalistic Evolution* (NE)

   Theology—no supernatural
   Teleology—no purpose
   Intervention—no intervention
   Genealogy—common descent
   Cosmology—old universe
   Process—spontaneous natural processes only

2. *Non-teleological Evolution* (NTE)

   Theology—Creator
   Teleology—no purpose
   Intervention—no intervention
   Genealogy—common descent
   Cosmology—old universe
   Process—conditions necessary for life established at creation

3. *Planned Evolution* (PE)

   Theology—Creator
   Teleology—purpose
   Intervention—no intervention

1. Rau, Mapping the Origins Debate, 41.

      Genealogy—common descent
      Cosmology—old universe
      Process—perfect creation naturally fulfills God's purposes

4. *Directed Evolution* (DE)

      Theology—Creator
      Teleology—purpose
      Intervention—intervention
      Genealogy—common descent
      Cosmology—old universe
      Process—changes in universe and life subtly directed over time

5. *Old-Earth Creation* (OEC)

      Theology—Creator
      Teleology—purpose
      Intervention—intervention
      Genealogy—de novo creation
      Cosmology—old universe
      Process—major body plans created over millions of years

6. *Young-Earth Creation* (YEC)

      Theology—Creator
      Teleology—purpose
      Intervention—intervention
      Genealogy—de novo creation
      Cosmology—recent creation
      Process—each "kind" created in one week, within the last 10,000 years

     In the original table Rau had listed the six models as a top row in a table and then listed the characteristics of each in the columns underneath each model. He started with naturalistic evolution on the left and ended with young-earth Creation on the right. So as you read from left to right you were seeing increasing degrees of supernatural intervention in the creation process as you examined each model. We started our discussion of Chapter 10 Macroevolution with a diagram based on the top row of this table. However it is worthwhile to explore how a person's worldview and bias can show up even in a simple table or graphic. Let's take the top row.

Naturalistic Evolution	Nonteleological Evolution	Planned Evolution	Directed Evolution	Old Earth Creation	Young Earth Creation

Abbreviate the titles.

NE	NTE	PE	DE	OEC	YEC

Now let's vary as a typical Christian might, or as reflects the average American population.

Or how would an evolutionist writing against creationism draw the diagram?

Or how would a professor in the 1950s draw the diagram?

All of these diagrams reflect the biases of the diagram's creator; some changes may be very intentional; other changes might have been completely unknown to the diagram creator. For example, not a single diagram above includes the possibility of a worldview that combines a young earth with an old universe, i.e., some combination of YEC and OEC. It is also possible that the author had no hidden agenda at all, and chose the categories because they were the only ones they could think of, and needed to have equal spacing so that the titles would fit and look nice. You may say this is ridiculous, and it is a bit of overkill. But be forewarned; even scientific diagrams and tables can be biased without saying a word.

In spite of these warnings, the data from Rau's Table 2.1 does help sort out where various people are coming from and his book gives an excellent description of the presuppositions of various positions and how it affects their theology. But beware, people are not so simple that we can pigeon-hole them into a specific category. Personally I feel that there are as many origins categories as there are people who hold them. Categories and diagrams are useful, as long as you are willing to modify it to fit the next exception you will surely meet, and don't assume that everyone's theology and origins perspective is coherent enough to even fit in a specific category. We are all works in progress, and only those with an axe to grind, money to make, a reputation to keep, a career to preserve, or a salvation to save, are going to never change; at least a little bit. It behooves us to grant others the same grace and courtesy that we would wish be extended to ourselves in our own thinking. That doesn't mean we abandon our core convictions; but please can we be gracious in our disagreement with others? Isn't that what Jesus did until he was forced to do otherwise?

"Ellie, why are you a young earth creationist? The data is heavily against you," Bert, and Ellie were sitting in the student union, wasting time between classes. There was a sprinkling of people around, but they had the eating area to themselves.

"Why do you say heavily? I agree the geologic data is against my convictions, but I would not call it heavy, I have some arguments for my side," Ellie looked quizzically at Bert.

"But you just sit there through Dr. Delzer's lectures as he goes off on an OEC lecture rant?"

"Bert, I enjoy hearing what he has to say. I don't agree with him regarding the age of the earth and how he handles the fall of man, but he has a lot to say that I find very useful, even if we disagree on a couple of areas." Ellie smiled. "Bert, you ought to know that, you are a philosophy major right?"

Bret grinned, "You're right. But darn it, do you know how many times I have been told by a YEC convert that I am on a slippery slope, and that my beliefs are going to cause me to go to perdition?

Ellie was laughing, "Probably way too many times right!"

"Well not that many times, but enough! Why do people do this anyway? Bert was laughing also.

"Well you know the answer to that Mr. Philosophy! You tell me!"

Bert was rolling his eyes, "Okay I'll try. Perhaps because they are insecure in their belief, not so much in their YEC, but in their understanding of what it really means to be Christian."

Ellie was laughing again, "I like it, keep going; you might be on to something."

Smiling Bert started in, "Well if you don't have a clear understanding of just what the Christian faith is, and are not able to distinguish the non-essentials of Christianity from the essentials, it is tempting to grab a minor item that means a lot to you and turn it into an essential. Then if another Christian attacks that non-essential it becomes very threatening. I have seen this go the other way also. I was visiting a small Christian university near here. Wanted to hear the speaker they brought in. He worked at a well-known research university in physics most of his life. He had become a Christian recently, so he was on the road with a Science and Christianity lecture."

"Was he YEC?"

"No! This speaker was a definite OEC or something else. Jesus was wonderful; but science as he was taught it was the way God did things, like it or not. This poor girl came up to him with some kind of YEC oriented question after his talk, and before the faculty could get there and intervene, he was screaming at her, "Come out of this, Come out of these lies!" It was terrible. I hope he didn't scar her for life!"

"That is pretty terrible, I shouldn't laugh, but it is kind of funny in a way," Ellie was appalled but still giggling. "Would you ever do that to me Bert?"

"If I did that, you would just laugh in my face, and I would flat deserve it!"

"You are right Bert; that is exactly what I would do."

Bert grinned, "You don't fool me Ellie. Behind that innocent girl face lays a mind like a steel trap. Speaking of which, what is going on with Jon these days?"

"Well I gave him my copies of *The New Testament Documents Are They Reliable* by F. F. Bruce, and *Basic Christianity* by John Stott. Knowing him, he'll have them read by the weekend."

"Has he read the New Testament or any of the Bible?" asked Bert.

"Almost all of it."

Bert replied back, "Has anything impressed him?" Gotten his attention?"

"I don't know for sure, he jokes around so much. But he was definitely impressed with F. F. Bruce's arguments for the reliability of the New Testament manuscripts." Ellie looked up at Bert.

Bert nodded, "I don't blame him, over 4000 ancient manuscripts that correlate to a huge degree, written within 100 to 300 years of the time it happened. Compare this to the 2 to 6 manuscripts that made it through the medieval period; these manuscripts give all we know about Plato and Socrates. And that, some 800 years after these men lived. The New Testament manuscripts are the most authenticated ancient manuscripts in existence, nothing else comes close. You can question what they wrote, but it is very hard to question that what we have is not what they wrote."

"That was what impressed him," replied Ellie.

"What about you Ellie?"

"What do you mean Bert?" Ellie quickly stared Bert in the face.

"What about you and Jon? Missionary dating is dicey business." Bert glanced down, knowing he had plunged over the edge.

"Bert, I am not dating him, I am not that stupid. At least I don't think I am . . ."

"You have no guarantee how this is going to pan out." Bert said this silently wishing he had not started the conversation. "Watch your heart Ellie, don't get involved in something you later wish you had never started."

"I know that Bert."

"I should not have said anything, it isn't my business," Bert was looking out the window at something.

"You are right it isn't; but I am glad you did anyway," Ellie was looking out the window also. "It's Katrina! Hello Katrina!" And then she started waving her arms frantically out the window towards the approaching Katrina. Katrina obviously could not hear her, but the arm waving did catch her attention, and within a few minutes she was inside sitting at the table with Bert and Ellie.

"Ellie! You are not going to believe who I saw walking holding hands together!" Katrina was a bit breathless from the walk into the student union.

"Girls!" murmured a smug Bert.

"Who? Do I know them?" gasped an anxious grinning Ellie.

"Dr. Swartzmann," exclaimed Katrina.

"Him! That idiot! With who? What happened to his live-in girlfriend Charlotte?" Ellie was a bit disgusted.

"Oh, they broke up last year, they say Dr. Swartzmann has been a pill ever since; that is until a month or so ago," Katrina was into this bit of juicy gossip. "You will never guess who he is with now!"

"Who?"

"who?"

"Mary, our Mary. Some say she has moved in with him." Katrina had a wicked little grin on her face.

"What!!" This wasn't coming from Ellie, but Bert who was slowly turning white as a sheet.

Katrina gasped as it slowly dawned on her what she had just done. "Oh my God, Bert, I did not know! I am so sorry! I did not know! They say Dr. Swartzmann is changing, he actually said something nice about religion the other day. I am so stupid, I better go!"

Bert was very quiet, and very shook up, "Maybe you had better go."

Katrina was starting to sob, "I am so stupid."

"He is still an arrogant jerk, right"

"I am afraid so," Katrina was almost as upset as Bert. "I had better go. I have made a fool of myself today."

"Well I might as well leave with you," stated a very solemn Bert.

"I don't want to talk about it anymore!" Katrina looked frightened.

"Neither do I. There are better things to talk about," Bert, Ellie and Katrina pushed their chairs back and gathered their things. The conversation was muted as they left, but it did occur.

What is a miracle? What is a divine intervention? If you are a philosophical materialist, it is completely forbidden. To budge even an inch means your worldview is possibly flawed to the core. Thus no quarter is ever granted in those circles. To those who feel like they have experienced one, it's fairly obvious; to others it is not so easy to explain. The scientist who is a Christian can find themselves trapped between two worlds. The present world of science, where the mere mention of a divine intervention will get you hooted out of a scientific meeting; and the church world where prayers are answered in unusual circumstances, and discussions of Jesus's miracles, death and resurrection abound. Possibly the only place a scientist can discuss the possibility of a divine intervention in the natural world as a part of a legitimate scientific investigation is among Christians, who are scientists, who are open to such. In this day and age that is not a given.

Many Christian scientists deal with the tension by compartmentalizing their lives. In scientific matters divine interruptions don't happen, but in everyday religious life they do; especially when we venture back in history to the time of Christ. This sometimes shows as a rejection of miracles after the time of Christ and the Apostles. Sometimes scientists, theologians, and academicians deal with the materialistic spirit of our age by resorting to deism, or various forms of a tempered type of deism. We invent constraints or "choices" God has made in how he interacts with our world. God has chosen to reveal himself in the natural world by using methods that can only be investigated by methodological naturalism. Thus intelligent design or any type of supernatural intervention is ruled out by fiat. But why do we make that choice "for God"? Are we afraid that the inability to scientifically justify some kind of divine intervention will make the concept of God illegitimate?

Notice what the last question implies. 1) If science cannot justify or prove God then He doesn't exist. Really? Since when did science become the only valid way to determine whether a statement is true or not? 2) If I or my enlightened group of academicians cannot come up with a way to prove God's intervention, then it does not exist. Are we the sum total of who gets to judge whether an intervention has occurred or not? Who are we? 3) Do we even know what a divine intervention is; and would we recognize it, even if it smashed us in the face? Do we realize how incredibly arrogant we sound when discussion of this sort comes out of our mouths?

What most people do not realize is how brief and short, western man's fascination with philosophy materialism has been. Science did not grow out of a move to philosophical materialism; it grew in spite of it. When William Whewell proposed the term scientist in 1833, the idea that you would have to lay your divine interventions at the door of the scientific meeting would have been greeted with howls of laughter from many. And science managed just fine, thank you, without having an enforced methodological naturalism put on all of its participants. Where do we get the idea that if the reins of science are loosened in this area, that disaster will ensue and all of science will collapse? Where is the historical precedent? An origins science that refuses to factor in the possibility of God's intervention, is going to be materialistic, and when it is all done, probably way off the mark, very boring and dry. I say that because the academic inventions of man become very boring with time, but the academic inventions of God never cease to amaze.

Jesus changes the water into wine at Cana—a miracle. Or—water evaporates from a lake rises into the air to become a cloud. The cloud transports the water above a vineyard where it rains and the rain soaks into the ground. All by means of a hydrologic cycle that appears to be unique in our known universe; whose origins are more mysterious than we care to admit. The grape vine plant soaks up the water through its roots and transports it up the branch into the grape. There the grape cells use its cellular mechanism, proteins, enzymes, and an entire tree of life's worth of un-explicable biological information to turn the water and soil nutrients into grape juice. The vineyard keeper with his marvelously made body, whose origins are shrouded in mystery, gathers the grapes and presses out the juice and ferments the juice into wine in his vats. There certain bacteria use the miracles of the cell to produce ethanol and digest the grape juice until its turns into wine. It is all a miracle just as much as Jesus turning the water in the vats into wine. The only difference is that we can partially explain how the latter miracle occurs, and use it and modify it to suit our own purposes as every modern wine factory does in its effort to produce the best wine. But no matter how much you explain away this latter miracle with scientific explanations, behind every facet that you explain and provide a mechanism for; there lurks deeper below the surface a deeper greater miracle that cannot be explained except by appealing to God. All you have really done is make the obvious miracle a little less accessible to the common man who does not have the scientific training. It is the true role of the scientist to explain that miracle in depth, for all to know, rather than hide it in mysterious technical jargon that only the scientist who is interested can understand. I often wonder if we scientists have done such a good job of hiding the miracle, that we have even hidden it from ourselves. Only the sin nature in man that desires to usurp God from His rightful place of glory would do such a thing. It is my hope that future scientists will not do this, but reveal that which was hidden from the dawn of the ages.

What is fine tuning? Is it a type of miracle, a divine intervention? Among theists it is certainly viewed as such. Materialists see them as incredibly unlikely coincidences, yet to be explained. We started with the fine tuning of the universe and its beginning; then moved on to the development of the universe so intelligent life was possible. Then we moved to galactic fine tuning, the privileged planet hypothesis, and the incredible set of circumstances and parameters it took to build our unique solar system. We finished with the history of the earth and moon, and the remarkable circumstances it took to make the earth habitable for life; plate tectonics, our clear atmosphere, etc.

My PhD thesis investigated the exchange and bulk movement of water hydrogen atoms in a frozen water crystal. In preparation, I had to learn a lot about the properties of water. What I found amazed me decades ago; and prompted further research into the "fine-tuned" properties of water. Michael Denton has summarized this in his new book *The Wonder of Water*[2] where Denton examines the unusual and unique role water plays in our existence. He was foreshadowed by the Harvard chemist Lawrence Henderson in his classic *The Fitness of the Environment* in 1913. This prompts a more general question: How far does fine-tuning go? How much of our world is uniquely adjusted for humans? Some of these questions were examined in detail in the early 1800s with the *Bridgewater Treatises*. Some of the examples examined in these volumes were overdone. Some examples have lasted to today, including the famous Paley's watch thesis. But in spite of past excesses the question can still be asked, how far does some kind of fine-tuning go?

Alan Chalmers raises the following interesting point as he finishes up his classic text, *What is this thing called Science?*

2. Denton, *The Wonder of Water*.

The outcome of all this can be summarized as follows. A wide range of laws within physics can be understood as causal laws. When this is possible, there is a ready answer to Boyle question concerning what it is that compels physical systems to behave in accordance with laws. It is the operation of the causal powers and capacities characterized by the laws that make systems obey them. However, we have seen that there are fundamental laws in physics that cannot be construed as causal laws. In these cases there is no ready answer to Boyles's question. What makes systems behave in accordance with the law of conservation of energy? I don't know. They just do. I am not entirely comfortable with this situation, but I don't see how it can be avoided.[3]

Robert Boyle's answer to this question was simple as Chalmers has noted; "God makes matter behave in accordance with the laws He has ordained[4]." Chalmers thinks this answer is inadequate but as seen in the previous quote, his own attempts to explain the non-causal laws, draws a complete blank. The causal laws are ones where a particle exerts a force on another particle according to a given mathematical relationship. The non-causal laws are ones where matter and energy are treated as bulk entities without regard to how one atom or molecule causes another reaction in another atom or molecule. One bit of matter pushing or forcing another bit of matter is not even regarded. The 1st and 2nd law of thermodynamics and other conservation type laws fall into this non-causal category. And it seems no one has a good idea as to why they exist and why they should be so. I will argue with Boyle, that God made the universe obey the way it does, so that we can exist and think about it all.

A few years ago I was introduced to a physical chemistry problem where the professor asked the student to calculate the barometric pressure on the summit on Mount Everest (8,848 meters or 29,028.9 feet). All you need was a standard barometric formula taught in physical chemistry classes and knowledge that the pressure at sea level 0 meters or 0 feet is 760 torr or 1 atmosphere. Don't worry; I will not make you go through the problem or bore you with needless calculations. But I was curious and went online to find what the atmospheric pressure really is on top of Mount Everest. It turns out to be a complicated problem. The actual air pressure on top of Mount Everest is slightly different from what the barometric formula will predict. And that difference makes a world of difference for a man attempting to climb Mount Everest without supplement oxygen to breathe. I

3. Chalmers, What is this thing called Science?, 225.
4. Chalmers, What is this thing called Science?, 214.

stumbled on the article in the *Journal of Applied of Physiology*, "Barometric pressures at extreme altitudes on Mt. Everest: physiological significance" by West, Lahiri, Maret, Peters, and Pizzo[5]. In this article the authors show that standard tables for the drop in air pressure as you go up in elevation predict an air pressure of 236 torr or 0.3105 atmospheres. The air pressure drops over 2/3's to just 31.05 percent of sea level pressure. Extensive study has shown at that level humans do not get enough oxygen to move sufficiently and shortly die. On October 24, 1981, some of the authors climbed Mount Everest, and measured the summit value at 253 torr[6] or just 33.29 percent of sea level pressure. That 2.24 percent difference is the only reason humans can barely climb Mount Everest without supplemental oxygen. In the article they quote: "Nevertheless, Messner and Habeler achieved the ultimate in 1978, reaching the summit of Mt. Everest (alt 8,848m) without supplementary O2. Their accounts (9, 13) clearly indicate the slim margin of reserve at these extreme altitudes."

The article explains this 2.24% difference higher pressure (17 torr higher pressure) is a result of Messner and Hableler climbing in late fall when the weather clears and the average barometric pressure on top of Mount Everest is near its maximum. In the summer the air pressure in the equatorial regions of the earth increase by 11.5 torr or 1.51 percent because of a "combination of complex radiative and convective phenomena." Summarizing, if Mount Everest were ~20 degrees further north in latitude, or if humans attempted to climb Mount Everest without supplemental oxygen in the winter, or early spring months they could not do it. It is also very interesting that pressure maximum that makes it barely possible to climb Mount Everest without supplemental oxygen also coincides with the only two clear months of the year where climbers dare attempt a summit of Everest. Why is it that humans are just barely able to reach the highest land elevation on earth breathing ambient air? Why are our bodies at their very best just able to reach this point?

The critic will argue, "Come on! Do you expect me to believe that man is fine-tuned physiologically to just be able to explore all the earth? If that is so why can't he dive by holding his breath to the bottom of the deepest ocean, the Mariana trench?" Valid point: maybe it is just an unusual coincidence that mankind can just reach the top of the highest peak on earth on his own. If it was 100 feet higher, he could not do it, a little further north, he could not do it; the climbing season and barometric high season match

5. West, "Barometric pressures at extreme altitudes on Mt. Everest," 1188–1194.

6. West, "Barometric pressures on Mt. Everest." 1062. 1999 measurements confirmed the initial 1983 results.

up, otherwise he could not do it. But am I alone in my amazement of this coincidence? Consider the last two sentences written by the article authors, and printed with the consent of the journal editors.

> Thus it is remarkable that the latitude-dependent increase in barometric pressure shown in Fig. 4 makes it just possible. It remains for someone to elucidate the evolutionary processes responsible for man being just able to reach the highest point on Earth while breathing ambient air.[7]

So how far does fine-tuning go in our nature world? Perhaps all the way down. As long as we can discover new things about our universe, world and selves, we will probably stumble on new levels of fine-tuning. I cannot justify this scientifically, but it is consistent with a view of God who wishes to leave signposts to His existence, but never in such a fashion to compel belief; but rather encourage belief as a volitional choice of the believer. So far the data seems to be leaning heavily in that direction.

To the functional materialist, fine-tuning, purpose (teleology), non-objective, and intelligent causes in origins are anathema. It militates against their worldview. They have little use for amazement and pondering the odd. Does backing up to an earlier more teleological friendly version of science harm the advancement of knowledge? I argue strongly that it does not. Indeed to continue our present philosophical materialist oriented course will prove devastating to science and the pursuit of knowledge in general. Why? Michael Polanyi provides a starting answer.

Michael Polanyi fled Germany to England and the University of Manchester to escape Hitler. There he abandoned his scientific studies and turned his extraordinary mental powers to the problem of why Europe destroyed itself with the disasters of WWI, WWII, and communism. He blamed it on the philosophical moral inversion that takes place in society when one tries to explain all knowledge through a lens of exact "objective" science; that reduces all moral values and religion with a reductionist, "objective," scientific epistemology. His explanation of the key role tacit knowledge plays in science was his underpinning, and his weapon to demonstrate the absurdity of any objective science without huge tacit knowledge components. Polanyi claimed that tacit knowledge undergirds all knowledge, and because of its un-specifiable from-to character could never be objectively formulated and described. In fact, all such attempts to do so would pervert the knowledge gained from such a tacit knowledge base, and lead to grievous errors in our pursuit of knowledge and truth.

---

7. West, "Barometric pressures at extreme altitudes on Mt. Everest," 1194.

As discussed earlier in Chapters 3 The Society of Science and Chapter 7 Science and Metaphors, tacit knowledge is a from-to knowledge where we concentrate on a particular focal point by seeing through the often subliminal particulars that make up the means by which we understand or perform the integrated whole we are focusing on. Tacit knowledge can be seen in the following examples:

- Line drawing of a staircase (See Chapter 2 Philosophy First)

- Recognizing a face

- Reading x-ray photographs

- Recognizing your Mother's voice

- Sight reading music

- How to put together a good experiment

- Dreaming up or creating a good hypothesis to test

- Recognizing when an experimental result is bad

- Recognizing when a particular theory is bad and should be abandoned

- Recognizing when a particular theory is good and should be kept— even when most of the evidence is against it. (Examples of scientific theories that initially had insufficient evidence to support them: Plate tectonics, heliocentric theory of solar system)

At this point I shall assume that tacit knowledge is understood, and that it undergirds all forms of empirical objective knowledge. Take a piece of empirical knowledge, undergirding it is a bit of tacit knowledge that was used to find or "discover" our empirical knowledge. As we build our empirical knowledge base we often use newly acquired tacit knowledge skills to find the new "objective" knowledge. For example when I mentored with my PhD advisor in the field of molecular spectroscopy, I acquired new tacit knowledge skills in interpreting molecular spectra that were essential to producing clear objective results. Interpreting molecular spectra often requires a type of connoisseurship that leads to one explanation of the spectrum rather than another. The choice of the next experiment to run to objectively confirm one's scientific suspicions is not given by clear objective rules, but often by an un-specifiable sense that, this rather than that, experiment will give us the confirmation we need. Great scientists build their careers and skills on this type of connoisseurship, and develop it greatly as they increase the objective knowledge base. Graphically we might illustrate this as follow[8].

8. Collier, "The Necessity of a Polanyian Perspective in the Science-Christianity

Relation of tacit knowledge to objective knowledge as knowledge grows

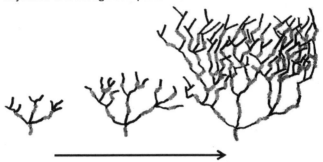

- Tacit knowledge component of field
- Objective knowledge component of field

Time/growth of knowledge tree

Since knowledge is growing and expanding as a branching tree might represent, we see that the un-specifiable tacit knowledge components are also growing in an exponential fashion also. If this model of knowledge is correct then we will forever have huge components of un-specifiable tacit knowledge undergirding our knowledge that will be forever beyond specific articulation, and thus the theory of everything where everything is known remains un-specifiable, and practically unreachable. This also illustrates why God of the Gap arguments are basically non-useful rhetoric. The un-specifiable tacit component gaps in our knowledge are not decreasing and explained by science, but are growing and increasing. It has to for our knowledge base to increase. Thus from a Polanyian perspective a God of the gaps argument is incoherent.

Polanyi raised another objection to the theory of everything, or a grand unification type of scientific perspective. Let us use his words from *Meanings*, 1975.

> Ever since Laplace first raised the point in defining Universal Knowledge, philosophers have discussed the notion that from today's topography of the ultimate particles of an object we can predict, by the laws of mechanics, any future topography of these particles. The immense difficulty of carrying out such computations is easily perceived. This has diverted attention from the far greater difficulty involved in the idea itself, namely, that the results of such a calculation would in themselves tell us

Dialogue."

nothing of any importance . . . We have seen that living beings are characterized by their physiognomies, including the space-time physiognomy of their functions. A physiognomy obeys no mathematical formula; it can be recognized only tacitly by dwelling in its numberless particulars, many of them subliminal. We can know a frog only by dwelling in its particulars in this way, and we can know the topography of a frog only if we are able first to know a frog . . . To attribute such levels of existence as these to an atomic topography seems as absurd as it would be to talk about the smell of differential equations, yet the modern mind seems hardly to hesitate in countenancing such incongruities. Cowed by the experience of the Copernican revolution, we dare not trust the testimony of our senses to contradict the teachings of science.[9]

Let us modernize this a bit. The future gurus and stars of science have ground out the final equation that predicts the wavefunction and motion of all particles in the universe. *(In quantum mechanics if you have the correct mathematical wavefunction, "function," for a particle then you theoretically can calculate any property you want about that particle).* The big blue computer of the future merrily crunches out the space-time coordinates of everything and the wavefunction at any particular point in space-time. Charlene zooms her space-time console to the planet Bart in a galaxy far, far away. Not knowing that she has discovered a planet with living things on it she zooms in on a small 3 by 3 by 3 inch cube of space. There she sees the space time coordinates of a particular location of carbon, hydrogen, nitrogen, oxygen, sulfur, iron, phosphorous, calcium and selenium atoms. Being a biochemist who has stared at thousands of electron density maps of DNA, she recognizes DNA. From there she quickly recognizes the electron density and wavefunction map of the alpha-helix of a protein, and then on to various tissues. Being a good biologist who has dissected thousands of specimens, and a good microscopist, she recognizes the various tissues of a living organism. However she also is a pretty good zoologist who specialized in aquatic animals, particularly frogs, and she knows how frogs jump and behave. "A frog," she gasps, "about to jump!"

But alas, our scientist Charlene is wrong, but still very close and about to discover a new type of animal. It is a hairless Bartian Sidthhog about to propel himself backwards by belching a large quantity of CO_2 out of its mouth. Unfortunately, our scientist Charlene is not the one at the controls. It is Sam, the very mathematical Calahari bushman running the console, and he has never even seen a frog; thus he "sees" nothing, and moves the

9. Polanyi, Meanings, 143.

controls to something more interesting . . . The map is not the thing. The thing is not the map. The theory of everything is but a map. The very act of reading a topographical map to "see" a valley and thus navigate ones way through that valley in reality is a tacit skill that is learned by mentoring, examples, practice and a base of tacit reading and drawing skills that boggle the imagination. To use a theory of everything to do anything; requires 4000 years of human apprenticeship, connoisseurship, and tacit knowledge skills; that so dwarf the equation, that to apply the name "theory of everything" to it, is absurd. From Polanyi:

> We are approaching here a crucial question. The declared aim of modern science is to establish a strictly detached, objective knowledge. Any falling short of this ideal is accepted only as a temporary imperfection, which we must aim at eliminating. But suppose that tacit thought forms an indispensable part of all knowledge, then the ideal of eliminating all personal elements of knowledge would, in effect, aim at the destruction of all knowledge. The ideal of exact science would turn out to be fundamentally misleading and possibly a source of devastating fallacies.[10]

But what are these fallacies Polanyi is talking about? How will it affect us if we buy the objectivist agenda? Again from Polanyi' classic work *Personal Knowledge*,

> He is strong, noble and wonderful so long as he fears the voices of this firmament *(the mutually recognized superior knowledge of others—italics explanation added)*; but he dissolves their power over himself and his own powers gained through obeying them, if he turns back and examines what he respects in a detached manner.. then law is no more than what the courts will decide, art but an emollient of nerves, morality but a convention, tradition but an inertia, God but a psychological necessity. Then man dominates a world in which he himself does not exist. For with his obligations he has lost his voice and his hope, and been left behind meaningless to himself.[11]

Michael Polanyi was a human, an extraordinarily bright one. He married, raised two children, and worried if his son John Polanyi would ever make it in chemistry. He waffled on his Christian beliefs when asked directly, did not make any strong connection with a particular church, and

10. Polanyi, The Tacit Dimension, 20.

11. Polanyi, Personal Knowledge, 380.

made many of the mistakes that all of us make. Eventually he grew old and slowly lost his ability to reason and think in the remarkable way that his readers knew and admired. In 1976 he died old and full of years hoping to see God the Father, God the Son, and God the Holy Spirit. He tried to speak to an academic world that was enraptured with positivistic science, and stop what he perceived as the final outcome of traditionless, valueless, objectivist thinking: disaster.

His world was the world of modernity, where the seeds of post-modernism were just being laid. Today our world is becoming post-modern. Do his words still ring true today? I will argue yes; more than ever. Post-modernism is really just hyper-modernism. The roots are the same. Post-modernism is not as positivistic as modernity and not as concerned whether the grand materialistic meta-narrative is true or not. But it still embraces a radical skepticism with regard to traditional human values that western civilization esteems; values given by Judeo-Christian ethics, and the implications on what it means to be thinking and human. Polanyi addressed his ideas towards this root skepticism and the flawed epistemology that spawned it. A Polanyian perspective, if engaged correctly, can help reset imbalance in the science-faith dialogue because nothing stops an intellectual bully or helps a truly enquiring skeptic as does, a very good question.

Anecdote—a narrative, usually brief, of a separable incident or event of curious interest, often biographical.

Anecdotal—Characteristic of or containing anecdotes.[12]

Many years ago when my kids were small, my wife and I took them to public libraries for story readings, and other events of interest to small children. This time we went north of our suburban home to a county library where a young American Indian couple was going to demonstrate Indian fancy dancing and Indian story-telling. The husband who was Crow, started first. Decked out in full American Crow Regalia he turned on the music of chanting, drumbeats, and then showed the sizable group of kids sitting on the floor, the various dances of the Crow tribe. He was a hit with the kids and parents. Then his young Caddo wife rose to tell an Indian story. Her name was Cricket.

"Once upon a time long, long ago, there was a Caddo chief and his wife. They were getting old and had no children. So they prayed to the Great Spirit and asked him to give them children. In time the wife became pregnant and gave birth to a son. But the son changed from a human and grew

12. Webster's Collegiate Dictionary.

up to become a very bad monster. This monster went around and killed a lot of the Caddo people. The chief and his wife were very sad, and asked the Great Spirit to save them and the Caddo people from this horrible monster. So the Great Spirit made it start to rain. It was raining so much that the water was rising up everywhere. The Great Spirit made a great big hollow reed, set it on the ground, and told the Indian chief and his wife and tribe to go inside of the huge reed. They did as they were told and the waters rose and rose as it continued to rain. The reed with the Caddo people was floating on top of the waters as they covered the earth. The monster was drowned but the Caddo people saved. Eventually the water receded, and when the Caddo chief and his people came out of the reed they found corn growing out of the ground. That is how the Great Spirit gave corn to the Caddo people!" with that Cricket finished her story while the kids and moms clapped and cheered. Then she turned to face the kids and asked, "What does that remind you of?"

"Noah and the Ark!" the kids yelled in partial unison.

"That is right! It sounds similar to Noah and the Ark," she replied back.

I waited until most of the crowd had dispersed and then wandered over to Cricket, "Hello Cricket, my name is Bill and I am here with my kids, and I just wanted to thank you and your husband for doing this for the kids."

"My pleasure Bill."

"What are you doing now?" as I suspected I was talking to a bright young lady.

"Oh, I am enrolled in the University of Oklahoma studying anthropology," she replied back.

"Can I ask you a question?" I tentatively queried.

"Sure, what is it?"

"Am I correct that most of the American Indian languages are related to each other?" that was my first question.

"Yes, that is right." She was gazing at me quizzically.

"Have they done linguistic studies based on the rate of divergence of two languages from each other as they change with time?"

She was staring me in the face, "Bill, I like the way you are thinking! Yes they have tried to estimate how far back it was to the common language from which all Indian languages were descended."

"How far back is that?"

"It is very hard to make that estimate with any real accuracy, but the estimates come in around 50,000 years ago. But you know what bothers me, the standard explanation is that the American Indian came from Asia through the Siberian Strait—Alaskan land bridge that they think existed around 20,000 years ago. But Bill, they have found South American Indian settlements near the southern tip of Argentina that date back to 20,000 years ago. How did they get there?" She had this look of puzzlement on her face.

"Can I ask another question?"

"Sure!"

"How many American Indian tribes have flood stories like the one you told," at that point I was thinking, "She is going to think I am a nut."

"Bill, have you heard of the Red Earth Indian Festival held here in Oklahoma every year?"

I replied, "I have never been to one, but I am familiar with it."

"Well, I am an official storyteller for my tribe. Every year the storytellers from different tribes get together at the Red Earth Festival, sit around a common campfire, and exchange stories. Well one year as we were sitting around the campfire, an Indian storyteller stood up and suggested that each tribe tell its flood story. As the night progressed each storyteller from each tribe stood up and told their flood story. They were all different but all had in common the great flood." Then Cricket leaned in closer to me, "Bill, there were over 50 different tribes present around that campfire, every single one of them had a flood story! How in the world? How in the world could this have happened!? Every tribe has a story! It had to have really happened!"

As Cricket was reliving the moment it was obvious that it was as surprising to her as it was to me. How in the world?

This is an anecdotal story. The trouble with anecdotal stories and evidence is that it is often based on hearsay. It is very difficult to prove or falsify, and it depends heavily on the truthfulness of the speakers and listeners with their often flawed perceptions. You bring such a story to a professional academic meeting at your own peril. It is dangerous to even bring it up in a neutral group of people. There will always be someone in the back who will stand up and say, "That is easy to explain, 20,000 years ago as the glaciers were melting, there were ice dams that broke and flooded huge swaths of land. These floods greatly impressed the primitive Indian inhabitants of that time and they passed these . . ." At that point you are tempted to interrupt and say,

"Yeah, Yeah, I have thought about that; and about the ancient possible breach of the Straits of Gibraltar that flooded and created the Mediterranean Sea; and when that Istanbul scientist told me the Black Sea used to be fresh and one third its present size, until a possible earthquake 20,000 years ago opened up the Bosporus Straits connecting the Black Sea to the Mediterranean Sea; and don't forget the geologist friend of mine who said geologically, floods leave very few deposits behind, (1/16th of an inch or less is typical for an average river flood); thus we could have a global flood and have no trace of its presence in the geologic record; so it could have happened anytime, and don't forget, . . . flood stories are everywhere in cultures all over the world, etc. . . . Some people just have to have a nice neat answer for everything, even if their supposed rationale has no more evidence in favor of it than any random musing I come up with . . ." Then there is the other person who says, "You know my pastor just finished a sermon series . . ."

But why bother. It is an anecdotal story that you probably should not have brought up in public. It convinces few. But you were there, and it would be a lie to say that it did not affect you. So you sit on it, ponder what it all means, and share it only with a few, very kindred spirits.

Jon and Ellie were sitting in the courtyard near the edge of campus. It was a bright, cool Sunday morning and in the distance a lonely church bell was tolling, calling its parishioners to worship.

"I love the sound of church bells in the distance. What do you think, Jon?"

"Not bad, doesn't compete with the huge church bells you hear in European cities, but they have been building those belfries for hundreds of years."

Ellie smiled, "You are right, it is hard to top the sound of those bells ringing through the countryside. How much history was behind the ringing of those bells? Peace, war, disaster, life, death, birth, all the great and terrible moments of life."

Jon turned to her, "Did you know that European churches still ring their bells at noon every day in honor of the Hungarian victory over the invading Moslems in 1456 AD at the Siege of Belgrade, Hungary?"

"That is a long time to ring bells."

"Yes it is," replied Jon.

"Where are you Jon? You have read everything I have ever given you about God and Jesus."

"Here is my favorite quote," He pulled out one of Ellie's books, flipped a few pages, and started reading.

> When people defend their world view, they are not defending reason, or God, or an abstract system: they are defending their own fragile sense of security and self-respect. It is as instinctive as defending one's body from attack. No one understood the psychology of this better than Kierkegaard. He recognized how subtly inter twined are our beliefs with our instinct for self-preservation, and counseled the greatest sensitivity for those who seek to lead someone from error into truth: First and foremost, no impatience . . . . A direct attack only strengthens a person in

his illusion, and at the same time embitters him. There is nothing that requires such gentle handling as an illusion, if one wishes to dispel it. If anything prompts the prospective captive to set his will in opposition, all is lost. . . . The indirect method. . . , loving and serving the truth, arranges everything . . . , and then shyly withdraws (for love is always shy), so as not to witness the admission which he makes to himself alone before God, that he has lived hitherto in an illusion.[13]

"And you?" Ellie asked.

"I am in," with that, Jon stood up grabbed Ellie's hand, and stepped into the sunlight.

13. Taylor, The Myth of Certainty, 25–6. Quoting Karl Barth in The Word of God and the Word of Man.

Bibliography

Augustine. *The Confessions of St. Augustine*. Translated by Rex Warner, 1963. New York: Mentor, the New American Library, 1963.

Axe, Douglas D. "Estimating the Prevalence of Protein Sequences Adopting Functional Enzyme Folds." *Journal of Molecular Biology* 341 (2004) 1295–1315.

———. "Extreme functional sensitivity to conservative amino acid changes on enzyme exteriors." *Journal of Molecular Biology* 301 (2000) 585–95.

———. *Undeniable*. New York : HarperOne, HarperCollins, 2016.

Axe, Douglas D. and Ann K Gauger. "Model and Laboratory Demonstrations That Evolutionary Optimization Works Well Only If Preceded by Invention: Selection Itself Is Not Inventive," *BIO-Complexity* 2 (2015) 1–13.

Barrow, John D. and Frank J. Tipler. *The Anthropic Cosmological Principle*. Oxford: Oxford University Press, 1986.

Barton, David. "A Death Struggle between Two Civilizations." *Regent University Law Review*, 12 (2000) 3–56.

Behe, Michael. *Darwin's Black Box—The Biochemical Challenge to Evolution*. New York: The Free Press, 1996.

———. *The Edge of Evolution—The Search for the Limits of Darwinism*. New York: The Free Press, 2007.

Behe, Michael J. and David W. Snoke. "Simulating evolution by gene duplication of protein features that require multiple amino acid residues." *Protein Science* 13 (2004) 2651–2664.

Blank, Jennifer. "Hitchhiker's Guide to the Early Universe." *Science and Technology Review*, September 2002. https://str.llnl.gov/str/September02/Blank.html.

Bohlin, Ray. "An Unwanted Premiere!" Plano, Texas: Probe Ministries International. https://probe.org/the-privileged-planet/.

Boorstin, J. Daniel. *The Discoverers*, First Vintage Books Edition. New York: Random House, 1985.

Bowring, S. A., J. P. Grotzinger, C.E. Isachsen, A. H. Knoll, S. M. Pelechaty, and P. Kolosov. "Calibrating Rates of early Cambrian Evolution." *Science* 261 (1993) 1297.

Bray, P. G., Martin, R. E., Tilley, L., War, S. A., Kirk, K. and Fidock, D. A. "Defining the role of PfCRT in Plasmodium falciparum chloroquine resistance." *Mol. Microbiol.* 56 (2005) 323–333.

Brooks, J., and G. Shaw. *Origin and Development of Living Systems*. London and New York: Academic Press, 1973.

Chalmers, Alan F. *What is this thing called Science?* 3rd ed. St. Lucia, Queensland, Australia: University of Queensland Press, 1999.

Chen, J. Y. Chem, C. W. Li, Paul Chien, G.-Q.Zhou and Feng Gao. "Wen'an Biota: A Light Casting on the Precambrian World." Paper presented to *The Origin of Animal Body Plans and Their Fossil Records Conference*, Kunming, China, June 20–26, 1999, sponsored by the Early Life Research Center and the Chinese Academy of Sciences; Chien, Paul, J. Y. Chen, C. W. Li, and Frederick Leung, "SEM Observation of Precambrian Sponge Embryos from Southern China, Revealing Ultrastructures Including Yolk Granules, Secretion Granules, Cytoskeleton, and Nuclei," Paper presented to the *North American Paleontological Convention*, University of California, Berkeley, June 26-July 1, 2001.

Chesterton, G. K. "The Ethics of Elfland." *A G. K. Chesterton Anthology*. San Francisco: Ignatius Press, 1985.

Clodd, Edward. *Pioneers of Evolution from Thales to Huxley*. 1897, reprinted 1972 and in 2009 Charleston, South Carolina: Bibliolife, 2009.

Collier, William B. "The Necessity of a Polanyian Perspective in the Science-Christianity Dialogue." *2015 Annual Meeting of the American Scientific Affiliation*. Oral Roberts University, Tulsa, Oklahoma, July 24–27, 2015.

Collings, Richard G. *Random Designer—Created from Chaos to Connect with the Creator*. Browning Press, 2004.

Craig, William Lane. *The Kalam Cosmological Argument*. New York: Barnes and Noble, 1979.

Darwin, Charles. "To J. D. Hooker 1 February [1871]." Darwin Correspondence Project, Cambridge University Library, Cambridge. https://www.darwinproject.ac.uk/letter/DCP-LETT-7471.xml.

Darwin, Charles. *On the Origin of Species by Means of Natural Selection 6th ed.* London: John Murray, 1872.

Davies, Paul. *God and the New Physics*. New York: Simon and Schuster, 1983.

Davies, Paul. *The Origin of Life*. London: Penguin Books, 2003.

Debus, Allen G. *Man and Nature in the Renaissance*. Cambridge, UK: Cambridge University Press, 1978.

Denton, Michael. *Evolution: A Theory in Crisis—New Developments in Science Are Challenging Orthodox Darwinism*. Chevy Chase, Maryland: Adler and Adler, 1986.

Denton, Michael. *The Wonder of Water: Water's Profound Fitness for Life on Earth and Mankind*. Seattle, Washington: Discovery Institute, 2017.

Descrates, Rene. "Rules for the Direction of the Mind." *Great Books of the Western World*, vol. 31 Chicago: *Encyclopaedia Brittanica*, 1952. In Zeb B. Long and Douglas McMurry. *The Collapse of the Brass Heaven*. Grand Rapids, Michigan: Chosen Books, Baker, 1994.

DeWitt, Richard.*Worldviews: An Introduction to the History and Philosophy of Science*. Oxford: Blackwell, 2004.

Dodd, M. S., D. Papineau, T. Grenne, J. F. Slack, M. Rittner, F. Pirajno, J. O'Neil, C.T. S. Little. "Evidence for early life in Earth's oldest hydrothermal vent precipitates." *Nature 543 (2017) 60.*

Dolgin, Elie. "Rewriting Evolution." *Nature 486 (2012) 460–2.*

Dose, Klaus. "The Origin of Life: More Questions than Answers." *Interdisciplinary Science Reviews 13 (1988) 348.*

Dupree, A. Hunter. *Asa Gray: American Botanist, Friend of Darwin*. Cambridge, Massachusetts: Harvard University Press, Belknap, 1959.

Dyson, Freeman. *Disturbing the Universe.* New York: Harper and Row, 1979.

Eddington, Arthur. *The Nature of the Physical World.* Whitefish, Montana: Kessinger, 2004.

Eigen, Manfred. "Selforganization of matter and the evolution of biological macromolecules." *Naturwissenschaften 58 (1971) 465–523.*

Enuma Elish, The Epic of Creation. Translated by L.W. King from The Seven Tablets of Creation, London 1902. Internet Sacred Text Archive. http://www.sacred-texts.com/ane/enuma.htm.

Erwin, D. H., M. Laflamme, S. M. Tweedt, E. A. Sperling, D. Pisani, and K. J. Peterson. "The Cambrian Conundrum: Early Divergence and Later Ecological Success in the Early History of Animals." *Science* 334 (2011) 1091–1097.

Feyerabend, Paul K. *Against Method: Outline of an Anarchistic Theory of Knowledge.* London: New Left Books, 1975.

Feynmann, Richard P. *QED The Strange Theory of Light and Matter.* Princeton, New Jersey: Princeton University Press, 1985.

Flew, Anthony. *There is a God.* New York: HarperCollins, 2007.

Foote, Michael, and Stephen Jay Gould. "Sampling, Taxonomic Description, and Our Evolving Knowledge of Morphological Diversity." *Paleobiology* 23 (1997) 181–206..

Godfrey, Laurie, ed. *Scientist Confront Creationists.* New York: W. W. Norton, 1983.

Gomes, R., H. F. Levison, K. Tsiganis, and A. Morbidelli. "Origin of the cataclysmic Late Heavy Bombardment period of the terrestrial planets." *Nature,* 435 (2005) 466–69.

Gonzalez, Guillermo and Jay W. Richards. *The Privileged Planet.* Washington DC: Regnery, 2004.

Gura, Trisha. "Bones, Molecules or Both?" *Nature* 406 (2000) 230–3.

Hall, Stephen S. "Hidden Treasures in Junk DNA." Scientific American, October 1, 2012. https://www.scientificamerican.com/article/hidden-treasures-in-junk-dna/.

Heeren, Fred. *Show Me God,* Revised Edition. Wheeling, Illinois: Day Star, 2000.

Hobbes, Thomas. *Leviathan.* Renascence Editions. Eugene, Oregon: University of Oregon. 1999.

Holmes, Bob and James Randerson. "Creationism special: A sceptic's guide to intelligent design." *New Scientist,* July 9, 2005.

Hoyle, Fred. "The Universe: Past and Present Reflections." *Engineering and Science,* November, 1981.

Johnson, Donald E. *Programming of Life.* Sylacauga, Alabama: Big Mac, 2010. The Veeramachaneni, Stein, and Lewis references were taken from quotes on pages 26–27 in Johnson's *Programming of Life,* where he discusses the phenomena of overlapping genes in detail.

Johnson, Phillip. *Darwin on Trial,* 2ⁿᵈ Ed. Downers Grove, Illinois: InterVarsity, 1993. The quotes on page 50 are taken from Eldredge's book, *Time Frames,* see research notes *Darwin on Trial,*183.

Kanitz, Lori. Dr. Lori Kanitz was formerly Chair and Associate Professor of English at Oral Roberts University, Tulsa, Oklahoma; and is now Assistant Director of the Institute for Faith and Learning at Baylor University, Baylor, Texas. She has kindly granted permission to reproduce parts of her lecture in this manuscript. Dr. Kanitz was asked to teach a guest lecture in the early days of the honors seminar Philosophy of Science, and she quickly became a regular guest lecturer. The lecture became a classic among the students and faculty that heard it. Credit for the ideas and references used in this science and metaphor chapter should properly go to her. All errors and misinterpretations belong to the author. All references to Polanyi and the final page explaining Polanyi's moral inversion arguments and quotes were inserted by the author.

Kellisa, Manolis, Barbara Wold, Michael P. Snyder, Bradley E. Bernstein, Anshul Kundaje, Georgi K. Marinov, Lucas D. Ward, Ewan Birney, Gregory E. Crawford, Job Dekker, Ian Dunham, Laura L. Elnitski, Peggy J. Farnham, Elise A. Feingold, Mark Gerstein, Morgan C. Giddings, David M. Gilbert, Thomas R. Gingeras, Eric D. Green, Roderic Guigo, Tim Hubbard, Jim Kent, Jason D. Lieb, Richard M. Myers, Michael J. Pazin, Bing Ren, John A. Stamatoyannopoulos, Zhiping Weng, Kevin P. White, and Ross C. Hardison. "Defining functional DNA elements in the human genome." *Proceedings of the National Academy of Science* 111 (2014) 6131–6138.

Kettlewell, H. B. D. "Darwin's Missing Evidence." *Scientific American,* February 20, 2009, 148–153. The story referenced in the March 2009 column "*50, 100, 150 Years Ago*" was originally published in the March 1959 issue of Scientific American.

Koestler, Arthur Koestler. *Act of Creation.* New York: Dell, 1964. In Zeb B. Long, and Douglas McMurry. *The Collapse of the Brass Heaven.* Grand Rapids, Michigan: Chosen Books, Baker, 1994.

Kraft, Charles. "Worldview and Spiritual Power." *Chinese Leadership Conference on Church Renewal.* Spring 1987. School of World Mission, Fuller Theological Seminary Pasadena, California. Cited by Zeb B. Long and Douglas McMurry. *The Collapse of the Brass Heaven.* Grand Rapids, Michigan: Chosen Books, Baker, 1994, 63.

Kuhn, Thomas S. *The Structure of Scientific Revolutions,* 3rd ed. Chicago, Illinois: University of Chicago Press, 1996.

Lakoff, George and Mark Johnson. *Metaphors We Live By.* Chicago: Univ. of Chicago Press, 1980.

Lee, Michael S. Y. "Molecular Phylogenies Become Functional." *Trends in Ecology and Evolution* 14 (1999) 177–8.

Lewis, Clive S. *The Discarded Image.* Cambridge, UK: Cambridge University Press, 1964.

———. *Mere Christianity.* New York, New York: Macmillan, 1960.

———. *Miracles.* New York: MacMillan 1960.

———. "Religion without Dogma." *God in the Dock.* Grand Rapids, Michigan: Eerdmans, 1970.

Lewis, R., B. Parker, D. Gaffin, M. Hoefnagels, *Life,* 6th ed. New York: McGraw-Hill Science/Engineering/Math, 2006.

Lindberg, David C. *The Beginnings of Western Science,* 1st ed. Chicago, Illinois: The University of Chicago Press, 1992.

Locke, John. *An Essay Concerning Human Understanding.* Kitchener, Ontario: Batoche, 2001. ProQuest Ebrary. June 3, 2015.

Long, Zeb B. and Douglas McMurry. *The Collapse of the Brass Heaven.* Grand Rapids, Michigan: Chosen Books, Baker, 1994.

Luskin, Casey. "The Top Ten Scientific Problems with Biological and Chemical Evolution." In *More Than a Myth.* Paul Brown and Robert Stackpole, eds. British Columbia, Canada: Chartwell Press, 2014.

Macrae, Norman. *John Von Neumann: The Scientific Genius Who Pioneered the Modern Computer, Game Theory, Nuclear Deterrence and Much More.* New York: Pantheon Books, 1992.

Maotianshan Shales. Wikipedia. https://en.wikipedia.org/wiki/Maotianshan_Shales.

Mayr, E. *Populations, Species and Evolution.* Cambridge, Massachusetts: Harvard University Press, 1970.

McGrath, Alister E. *Science and Religion: An Introduction.* Oxford: Blackwell, 1999.

———. *Science and Religion—A New Introduction,* 2nd ed. West Sussex, UK: Blackwell-Wiley, 2010.

Meyer, Stephen Meyer. *Darwin's Doubt: The Explosive Origin of Animal Life and the Case for Intelligent Design.* New York: HarperOne, HarperCollins, 2013.

———. *Signature in the Cell: DNA and the Evidence for Intelligent Design.* New York: HarperCollins, 2009.

Meyer, Stephen C., Scott Minnich, Jonathan Moneymaker, Paul A. Nelson and Ralph Seelke. *Explore Evolution—The Arguments For and Against Neo-Darwinism.* Melbourne and London: Hill House, 2003.

Midgley, Mary. *Science and Poetry.* Routledge, 2000. eBrary, Oral Roberts University Library, Tulsa, February 5, 2007. http://site.ebrary.com/lib/oru.

Miller, Kenneth. "The Collapse of Intelligent Design: Section 5 Bacterial Flagellum" Archived 2016-10-17 at the Wayback Machine. Case Western Reserve University, January 3, 2006.

Mirsky, Steve. "In the beginning was the cautionary advisory." *Scientific American,* February 2005.

Moleski, Martin X. "Polanyi vs Kuhn: Worldviews Apart." The Polanyi Society, Missouri Western State University: http://www.missouriwestern.edu/orgs/polanyi/TAD%20WEB%20ARCHIVE/TAD33-2-fnl-pg8-24-pdf.pdf

Montaigne, M. *Essays,* ii, xii. Translated by E. J. Trechmann. 1927. London: Oxford University Press. As quoted in John D. Barrow and Frank J. Tipler, *The Anthropic Cosmological Principle,* Oxford: Oxford University Press, 1986.

Morbidelli, A., H. F. Levison, K. Tsiganis, and R. Gomes. "Chaotic capture of Jupiter's Trojan asteroids in the early Solar System." *Nature,* 435 (2005) 462–65.

Myers, P. Z. Myers quoted in. "Irreducible Complexity." RationalWiki. https://rationalwiki.org/wiki/Irreducible_complexity.

Neumann, John V., Arthur W. Burks, ed. *Theory of Self Reproducing Automata.* Urbana, Illinois: University of Illinois Press, 1966.

Overbye, Dennis. *Big Brain Theory: Have Cosmologists Lost Theirs?* New York Times, Science, January 15, 2008.

Owens, Virginia Stem. "Telling the Truth in Lies." *Encounters: Readings for Advanced Composition,* William R. Epperson and Mark R. Hall, eds. 9–12. Dubuque: Kendall/Hunt, 2001.

Parker, Eric T., Henderson J. Cleaves, Jason P. Dworkin, Daniel P. Glavin, Michael Callahan, Andrew Aubrey, Antonio Lazcano, and Jeffrey L. Bada. "*Primordial synthesis of amines and amino acids in a 1958 Miller H_2S-rich spark discharge experiment.*" *Proceedings of the National Academy of Science.* 108 (2011) 5526–31.

Pascal, Blaise. *Pensees,* II. 77. In Zeb B. Long and Douglas McMurry. *The Collapse of the Brass Heaven.* Grand Rapids, Michigan: Chosen, Baker, 1994.

Pearcy, Nancy R. and Charles B. Thaxton. *The Soul of Science.* Wheaton, Illinois: Crossway, 1994.

Petto, Andrew J. and Laurie R. Godfrey, eds. *Scientists Confront Creationists: Intelligent Design and Beyond.* New York: W. W. Norton, 2007.

Planck, Max. *Scientific Autobiography and other Papers.* Translated by F. Gaynor. 1949. New York.

Poirier, Maben Walter. "A Comment on Polanyi and Kuhn." *The Thomist* 53 (1980) 261.

Polanyi, Michael. "Life's Irreducible Structure." *Science* 160 (1968) 1308–12.

———. *Meanings.* Chicago, Illinois: Univ. of Chicago Press, 1975.

———. *Personal Knowledge.* Chicago, Illinois: University of Chicago Press, 1958.

———. *Science, Faith and Society.* Chicago, Illinois: University of Chicago Press, 1964, originally in 1946 by Oxford University Press.

———. *The Tacit Dimension.* Gloucester, Massachusetts: Peter Smith, 1983.

Postman, Neil. *The End of Education: Redefining the Value of School.* New York: Alfred A. Knopf, 1995.

Privileged Planet DVD. La Mirada, California: Illustra Media, 2005.

Rana, Fezale and Hugh Ross. *Who Was Adam? A Creation Model Approach to the Origin of Man.* Colorado Springs, Colorado: NavPress, 2005.

Rau, Gerald. *Mapping the Origins Debate.* Downers Grove, Illinois: InterVarsity, 2012.

Rensselaer Polytechnic Institute. "Earth's Early Atmosphere." *Astrobiology Magazine,* December 2, 2011. https://www.astrobio.net/geology/earths-early-atmosphere/.

Richards, Jay. *List of Fine Tuning Factors.* Seattle, Washington: Discovery Institute Website. http://www.discovery.org/f/11011.

Ross, Hugh Ross. *The Creator and the Cosmos,* 3rd ed. Colorado Springs, Colorado: NavPress, 2001; Covina, CA: Reasons to Believe. "Why the universe is the way it is." https://d4bgeo2xg5qba.cloudfront.net/files/compendium/compendium_part1.pdf.

Ross, Hugh. *Improbable Planet.* Grand Rapids, Michigan: Baker, 2016.

Russell, J. B. *Inventing the Flat Earth Myth Columbus and Modern Historians.* Westport, Conneticut: Praeger, 1991.

Schopf, J. William. *Cradle of Life.* Princeton, New Jersey: Princeton University Press.

Shapiro, Robert. *Origins: A Skeptic's Guide to the Creation of Life on Earth.* New York: Summit Books, 1986.

Soskice, Janet Martin. *Metaphor and Religious Language.* Oxford: Clarendon, 1987.

Stein, L. "Human Genome: End of the Beginning." *Nature* 431 (2004) 915–6.

Stevens, Will. "Self-Replicating Machines." The following diagrams are based loosely on the webpage ideas of Dr. Will Stevens at http://www.srm.org.uk/introduction.html.

Synder, Laura J. *The Philosophical Breakfast Club.* New York: Broadway, Crown, 2011.

Taylor, Daniel. *The Myth of Certainty.* Downers Grove, Illinois: InterVarsity, 1999. Quoting Karl Barth. *The Word of God and the Word of Man.* Gloucester, Massachusetts: Peter Smith, 130–1.

Thaxton, Charles B., Walter L. Bradley, and Roger L. Olsen. *The Mystery of Life's Origin.* Dallas, Texas: Lewis and Stanley, 1984.

Thurman, L. Duane. *How to Think About Evolution and Other Bible-Science Controversies.* Downers Grove, Illinois: InterVarsity, 1978.

Trail, D., E. B. Watson, and N. D. Tailby. "The oxidation state of Hadean magmas and implications for early Earth's atmosphere." *Nature* 480 (2011) 79–82.

Tsiganis, K., R. Gomes, A. Morbidelli, and H. F. Levison. "Origin of the orbital architecture of the giant planets of the Solar System." *Nature* 435 (2005) 459–61.

Valentine, J. W., and D. H. Erwin. "Interpreting Great Development Experiments: The Fossil Record." In *Development as an Evolutionary Process*. R. A. Raff and E. C. Raff, eds. 71–107. New York: Liss, 1987.

Veeramachaneni, V. W. Makalowskil, M. Galdzicki, R. Sood, and I. Makalowska. "Mammalian Overlapping Genes: The Comparative Perspective." *Genome Research* 14 (2004) 280–286.

Webster's Collegiate Dictionary, Fifth Edition, Springfield, Massachusetts: G. & C. Merriam, 1937.

West, John B. "Barometric pressures on Mt. Everest: new data and physiological significance." *Journal of Applied of Physiology*, 86 (1999) 1062.

West, John B., S. Lahiri, K. H. Maret, R. M. Peters, and C. J. Pizzo. "Barometric pressures at extreme altitudes on Mt. Everest: physiological significance." *Journal of Applied of Physiology*, 54 (1983) 1188–1194.

Yockey, Hubert P. "Origin of Life on Earth and Shannon's Theory of Communication." *Computers and Chemistry* 24 (2000) 105–23.